T0324056

An Introduction to Classical and Modal Logics

Classical logic—which studies the structural features of purported claims of fact—and modal logic—which studies relations of necessity and possibility—are different but complementary areas of logical thought. In this lively and accessible textbook, Adam Bjorndahl provides a comprehensive and unified introduction to the two subjects, treating them with the same level of rigour and detail and showing how they fit together. The core material appears in the main text, with hundreds of supplemental examples, comments, clarifications, and connections presented throughout in easy-to-read sidenotes, giving the book a distinct conversational feel. A detailed, multi-part appendix covers important background mathematical material that some students may lack, such as induction or the concept of countable infinity. A fully self-contained learning resource, this book will be ideal for one- to two-semester-long upper-level university courses on either or both of these topics.

ADAM BJORNDAHL is Associate Professor of Philosophy at Carnegie Mellon University. He has published numerous articles in journals including *The Review of Symbolic Logic*, *Res Philosophica*, and *Studia Logica*.

An Introduction to Classical and Modal Logics

The Outlines of Knowledge

ADAM BJORNDAHL

Carnegie Mellon University, Pennsylvania

CAMBRIDGE
UNIVERSITY PRESS

CAMBRIDGE
UNIVERSITY PRESS

Shaftesbury Road, Cambridge CB2 8EA, United Kingdom

One Liberty Plaza, 20th Floor, New York, NY 10006, USA

477 Williamstown Road, Port Melbourne, VIC 3207, Australia

314–321, 3rd Floor, Plot 3, Splendor Forum, Jasola District Centre,
New Delhi – 110025, India

103 Penang Road, #05–06/07, Visioncrest Commercial, Singapore 238467

Cambridge University Press is part of Cambridge University Press & Assessment,
a department of the University of Cambridge.

We share the University's mission to contribute to society through the pursuit of
education, learning and research at the highest international levels of excellence.

www.cambridge.org
Information on this title: www.cambridge.org/9781009450690

DOI: 10.1017/9781009450669

© Adam Bjorndahl 2024

This publication is in copyright. Subject to statutory exception and to the provisions
of relevant collective licensing agreements, no reproduction of any part may take
place without the written permission of Cambridge University Press & Assessment.

When citing this work, please include a reference to the DOI 10.1017/9781009450669

First published 2024

A catalogue record for this publication is available from the British Library.

Library of Congress Cataloging-in-Publication Data
Names: Bjorndahl, Adam, author.
Title: An introduction to classical and modal logics : the outlines of
      knowledge / Adam Bjorndahl.
Description: 1. | [New York] : Cambridge University Press, [2024] | Series:
      Cambridge introductions to philosophy | Includes index.
Identifiers: LCCN 2023053603 | ISBN 9781009450690 (hardback) | ISBN
      9781009450676 (paperback) | ISBN 9781009450669 (ebook)
Subjects: LCSH: Logic – Textbooks. | Modality (Logic) – Textbooks.
Classification: LCC BC108 .B53 2024 | DDC 160–dc23/eng/20240322
LC record available at https://lccn.loc.gov/2023053603

ISBN 978-1-009-45069-0 Hardback
ISBN 978-1-009-45067-6 Paperback

Cambridge University Press & Assessment has no responsibility for the persistence
or accuracy of URLs for external or third-party internet websites referred to in this
publication and does not guarantee that any content on such websites is, or will
remain, accurate or appropriate.

To all my students and all my teachers,
in every walk of life.

Contents

Figures

Tables

Preface

I began writing these pages in 2020, without knowing they would become a book. The pandemic had just descended, forcing classes around the world to suddenly switch to a remote format. I was teaching upper-year courses in Modal Logic and Game Theory at the time, to a combined total of over 100 students, and in both classes I made heavy use of the blackboard.

Zoom doesn't provide great substitutes for blackboards, and I eventually accepted that I would have to change the way I taught these subjects in a fairly substantial way.[1] Using a slide deck felt cruel—slides are good for presenting highlights, not details, and new concepts come and go too quickly. I needed something denser, a format that could support detailed proofs and worked examples. But was I really going to run my classes by reading through paragraphs of text on a shared screen?

Sidenotes saved me. A traditional textbook is too dense to lecture from directly, and crucially, too *linear*—in the classroom, I write the main ideas, definitions, proofs, and examples on the board, but I supplement all this with comments, asides, and answers to questions, interspersed throughout at irregular intervals. Sidenotes allowed me to create an approximation of this dynamic style of presentation in the static format of words-on-a-page.

So: as you read this book, you'll find the main ideas laid out in an orderly and efficient manner[2] in the main text. But in the *margins* of almost every page, you'll find the wider *conversation*—everything from alternative formulations

[1] I did purchase a handsome 36" × 48" board and I hung it in my living room (where my new "office" was), chalk dust notwithstanding. It remains there to this day. But it wasn't enough.

[2] I hope!

to irreverent remarks; from clarifications and reminders to broader contextualizations. This is not a traditional textbook. As much as possible, I have endeavoured to present the material in the same way I *teach* the material; in this spirit, my hope is that this book can be as much a resource for teaching as it is for learning.

The following semester was also, of course, fully remote, and I taught a course on classical logics in the same style. Week by week, I watched the course notes grow, and I began to realize that they could be combined into something broad and systematic yet lightweight and conversational; an introduction to logic for novices, but also a deeper dive for those who wish to hone their expertise. Something that could equally be read at home or lectured from in a classroom. The outlines of knowledge.

How to use this book

These pages contain roughly two semesters worth of material. Chapters 1 and 2 naturally form the backbone of a traditional "Introduction to Logic" class, with the option of covering additional material from the beginning of Chapter 3 to close out the semester. Chapter 1 starts from the very beginning and is designed to be friendly even for complete novices to the field of logic.

Chapter 3 forms the basis for an "Introduction to Modal Logic" class suitable for students who have some background in classical logic already under their belts.[3] In this context Chapters 1 and 2 provide review/reference material, while Chapters 4 and 5 offer options for exploring additional topics of interest beyond the core curriculum.

Appendix A collects a variety of background mathematical tools and techniques that students and readers may appreciate a refresher or overview of, from basic set notation to mathematical induction to notions of infinity. Readers who are approaching this topic as newcomers not only to logic but to mathematics as a whole ought to begin with Appendix A.1.

[3] But who may have never even heard the word "modal" before.

Perhaps the most important part of this book are the 159 exercises it contains. Mathematics is not a spectator sport; good, clear explanations are only useful to a point. Ultimately, one must learn by doing—there is no substitute for hitting dead ends and getting frustrated, making mistakes, re-reading (and re-re-reading) sections, asking for help, giving up, and going to bed only to jolt upright at 3 a.m. with the solution finally having materialized in your mind. Selected solutions to these exercises have been provided in Appendix B for reassurance and in case of emergency, but they come with this stern warning: the reader who jumps to the solutions without submitting themselves to the heartache described above thereby robs themselves of half the value of this book. You have been warned!

Acknowledgements

Writing acknowledgements is somewhat paralyzing in the prospect of mistakenly omitting some deserving name or committing some other grievous error. I can only hope that those I inadvertently yet inevitably offend in this manner will accept my only excuse, namely, that I am very, very bad at this.

From its inception to the moment I submitted the final draft, this book has benefitted from the inspiration, careful attention, and detailed feedback of numerous talented and generous individuals, including (in rough chronological order): Chris Grossack, Brad Bjorndahl, Chuong Truong, Peter Khoudary, Tandri Einarsson, Marina Dubova, Cynthia Wang, and Gavin Burke.

I am more broadly indebted to all the students I taught using (various stages in the development of) this manuscript. Thoroughly and repeatedly "playtesting" the material in this way was probably the single greatest contributor to whatever value ended up in these pages. While I can't list the names of all these students individually, I like to think that if any of them chance to flip through this volume, they will recognize the role they played in making it what it is.

I would also be remiss not to mention two people who played key roles in the production of the book itself. First, Juliana Arciniegas, who with grace, creativity, and skill crafted a beautiful cover image based on only my vaguest intimations; and second, Beverley Lawrence, whose attention to detail and perfectionism in the copy-editing process surpassed (to my utter delight) even my own.

Finally, I have been buoyed in this, my first book-writing quest, by the patience, friendship, and loving support of many people I have been very lucky to have in my life; too many to name, but a few I must highlight: Alejandra Arciniegas, Eloise Ballou, my wife Christina Bjorndahl, and of course my children, who make everything worthwhile—Anagnostis and Athena.

Introduction: What Is Logic?

Mathematics is the systematic study of patterns and structure. Number theory, for example, studies patterns and structure in numbers; geometry studies patterns and structure in space and figures; real analysis studies patterns and structure in real-valued functions.

What if we turned this focus inward, and engaged in the systematic study of mathematics itself? This is one way of understanding what logic is, and why it's reasonable to view logic as being both *a part of* mathematics while also in a certain sense *subsuming* mathematics. It is a part of mathematics in the sense that it studies patterns and structure in a certain domain, like all branches of math; it subsumes mathematics in the sense that its object of study includes the formalization of systematic reasoning, that is, math itself.

Introductory logic courses don't tend to emphasize this conceptual "inward looking" nature; instead, the subject is often introduced more directly as a tool for formalizing reasoning patterns like *modus ponens*[1] or the *law of contraposition*.[2] This, too, is a fair perspective on what logic is, and it can be quite far-reaching: sufficiently expressive logics can be used to reason about phenomena far outside the scope of traditional mathematics, such as the analysis of knowledge and belief, justification and evidence, obligation and permission, time, action, causation, learning, and much more.

[1] From "p" and "p implies q" we can infer "q".

[2] If "p implies q" then "not q implies not p".

In this book we take both views. We motivate the logical languages we focus on by their ability to formalize reasoning patterns of interest—both in and out of mathematics—and develop mathematically precise notions of *truth* for the various languages that we consider, leading to a rich array of *models* that support and extend our ability to reason. We also explore many "metalogical" results, that is, results about the logics themselves. The primary examples of this are *soundness and completeness* theorems, which (roughly speaking) tell us that we can only prove true things, and that we can prove *all* the true things, respectively. In other words, we'll spend a good portion of our time developing proofs *about proofs*.[3]

Formalizing reasoning

What does it mean to "formalize reasoning patterns", and what's the use of it? Let's return to the examples raised above: modus ponens probably seems pretty solid, but perhaps the law of contraposition looks a bit suspicious.[4] Is this just a flat assumption about how implication (and negation) works, or is it somehow derivable from more "basic" principles?

It depends. An important part of logic is the development of so-called "deduction systems", in which some statements, called *axioms*, are taken to be true by assumption, while others are derived from these axioms. It's certainly possible to adopt the law of contraposition as an axiom,[5] but we can also see how it follows from other principles. Consider the following, informal line of reasoning.

> Suppose p implies q. Then, if q is false, p must also be false, because if p were true it would imply that q is also true, which would be a contradiction!

This might feel convincing once you read it through a few times, but the project of logic is not merely to present rhetorically convincing arguments, but to establish the relationship between

[3] This enterprise leads to some of the deepest and most incredible results in mathematics, such as *Gödel's Incompleteness Theorem*, which says (roughly) that there are true statements of arithmetic that we cannot prove, and no matter how hard we work, we will never close this gap between truth and provability.

[4] Applying it to an example or two might make it feel more plausible—for example, since being a dog implies being an animal, it follows that if something is *not* an animal it's also not a dog. But an example is not the same thing as a proof!

[5] And many well-known axiom systems do just that.

assumptions and conclusions. The above is an attempt to justify the law of contraposition, but what are the assumptions we are making in the course of that justification? We can often get a clearer view by switching to a more formal presentation, something more reminiscent of a "proof".

1. *p* implies *q* (supposition)
    2. *q* is false (supposition)
        3. *p* is true (supposition)
        4. *q* is true (from 1 and 3)
        5. *contradiction* (from 2 and 4)
        6. *p* is false (from 3 and 5)
7. *q* is false implies *p* is false (from 2 and 6)

Laid out like this, we can identify several assumptions at play. For example, step 4 relies on modus ponens. Step 5 essentially labels anything that is both true and false a "contradiction", while step 6 relies on assuming that if you can derive a contradiction from a supposition, that supposition must be false. Step 7 relies on the idea that anything you can derive from a supposition is implied by that supposition.[6]

You might find all of these assumptions quite reasonable, even obvious, at an intuitive level. And intuitions about what's reasonable often guide the choices that mathematicians and logicians make for what counts as an axiom. On the other hand, reasonable people can disagree about what's obvious. Perhaps more concretely: which assumptions are appropriate typically depends on the context and/or application in mind.

For instance, though in some ways it might seem obvious that every statement must be either true or false,[7] there are contexts where this is not so clear. What if it's impossible to determine whether *p* is true or false? Does it still make sense to say that, nonetheless, it really *is* either one or the other? Dropping this assumption leads to an important "non-classical" logic called *intuitionistic logic*, closely related to the field of *constructive mathematics*.[8] Other examples of non-classical logics

[6] We are also making assumptions about the very structure of proof itself, like "You can make suppositions as part of a proof", "You can nest suppositions inside other suppositions", etc.

[7] As an axiom, this assumption is called the *law of the excluded middle*.

[8] Once you begin doubting the law of the excluded middle, the reasoning that supported step 6 in the proof sketch above might also start to seem a bit shaky.

[9] Allowing some statements to be both true and false.

[10] Admitting three or more truth values.

[11] Incorporating a notion of "relevance" to implications.

[12] For reasoning about knowledge, belief, evidence, and information ...

[13] ... about actions and events over time ...

[14] ... about obligations and permissions ...

[15] ... about what *would* have been the case if things had been different ...

[16] ... about the structure of arguments, formal and informal.

include *paraconsistent logic*,[9] *many-valued logics*,[10] *relevance logic*,[11] and *modal logics*, the latter of which are covered extensively in this book beginning in Chapter 3, and include *epistemic logics*,[12] *temporal and dynamic logics*,[13] *deontic logics*,[14] *counterfactual logics*,[15] and *justification logics*,[16] among many others.

Ultimately, the role of logic is not to arbitrate "The One True Truth", but rather, to provide a framework within which we can express and reason systematically about a variety of assumptions and their respective consequences. This book focusses on classical and modal logics. We begin with the former because of the foundational position they occupy in mathematics and philosophy, both historically and conceptually, and also because it makes good sense pedagogically to develop our tools in simpler settings before expanding and generalizing them.

1   Classical Propositional Logic

We begin with modest ambitions: to formalize those aspects of reasoning (mathematical or otherwise) that involve the basic concepts of *negation*,[1] *conjunction*,[2] *disjunction*,[3] and *implication*.[4]

We want to capture the essential logical structure of how these words function to join thoughts and sentences together. We approach this in several stages. The first, foundational stage[5] is to specify the *language* of propositional logic—that is, what kind of statements we are concerned with. This involves introducing official symbols that are meant to stand for the concepts above, and being very precise about how we are allowed to combine these symbols into what are called *formulas*.[6]

But symbols in and of themselves don't have *meaning*—not unless we somehow *give* them meaning. Thus, our next step[7] is to develop a mathematical framework that associates *truth* and *falsity* with the formulas of our language,[8] thus transforming them from simple marks on a page into something closer to what we normally think of as building blocks of thought and communication: statements that can be right or wrong. The meaning of "and", for example, is cashed out by specifying its *truth conditions*—that is, by telling you when a statement involving a conjunction like "*p* and *q*" counts as being true, namely, when *p* is true and *q* is also true.

After this[9] we approach the quest for meaning from a different, complementary angle: defining a notion of "deduction"[10] which essentially imbues formulas with meaning by telling us how we are allowed to *reason* with them. This may sound similar to the truth-based notion of meaning sketched in the previous

[1] "It is not the case that ..."
[2] "... and ..."
[3] "... or ..."
[4] "if ... then"

[5] Section 1.1.

[6] This is similar in spirit to specifying the *syntax* of a programming language—which strings of symbols count as proper code, and which will fail to parse.

[7] Section 1.2.
[8] Often called a "semantics".

[9] Section 1.4.

[10] Sometimes called a "proof system".

paragraph, but it is actually quite distinct: one can explain how to reason with formulas in the language without making direct reference to any notion of "truth". For instance, we might agree that from "p and q" one should be able to deduce "p".

[11] Section 1.5.

Finally, we conclude this chapter[11] by showing that these two attempts to give meaning to formulas are tightly connected: roughly speaking, the statements which are "always true" turn out to be precisely the statements that we can deduce through "pure reasoning". Naturally, all these high-level intuitions are made precise in the sections that follow.

1.1 Language

It may seem pedantic to carefully lay out which statements are "allowed" in propositional logic; after all, in the rest of mathematics, although we follow certain conventions about how to write things, we don't tend to *formally* specify these conventions. But this is exactly the point of logic—we want to study the practice of reasoning (mathematical or otherwise) with the same precision that mathematicians bring to the study of numbers, or functions, or sets. To do this we need to formally define our objects of study, which in this case are certain kinds of statements.

An analogy might be helpful here. In arithmetic, we consider expressions like "$2 + 2 = 4$", or "$8 \div 4 = 2$", or even "$3 \cdot 5 > 20$". Some are true and some are false, but at the level of syntax, they are all *well-formed*. By contrast, we do *not* consider expressions like "$2 + \div 3 = 5$" or "$18 >< 2 =$". These expressions are simply not allowed—they don't parse.[12]

[12] Another analogy, as mentioned in sidenote 6, is to writing a computer program—a line of code won't compile if it isn't written according to the official rules of the programming language.

The *formulas* of propositional logic are defined *recursively*. This means we start by specifying the most "basic" things that count as formulas, and then explain how to build new formulas from existing ones, step by step. Before presenting this in full generality, let's consider a concrete example. We might begin with two "basic" formulas, say named p and q, and then combine them to form the formula $p \wedge q$.[13] Then from this

[13] This "wedge" symbol stands for "and".

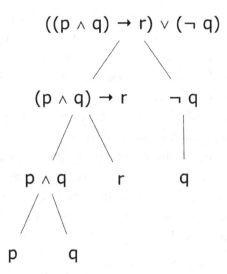

Figure 1.1 A visual representation of the construction
of the formula $((p \land q) \to r) \lor (\neg q)$.

formula and another "basic" formula r we could create $(p \land q) \to$ r.[14],[15] Separately, starting from q we could create $\neg q$;[16] we could then combine this formula with the one created before to create $((p \land q) \to r) \lor (\neg q)$.[17] This process can stop at any point and proceed in any order, leading to an infinite variety of possible formulas. The particular construction described in this paragraph is depicted pictorially in Figure 1.1.

Even more concretely, we might wish to reason about a statement like "if it's raining and I'm not carrying an umbrella, then I'm wet", which could be written as follows:

$$(raining \land \neg umbrella) \to wet.$$

Here we are treating "it's raining", "I'm carrying an umbrella", and "I'm wet" as "basic" statements, denoted by *raining*, *umbrella*, and *wet*, respectively. We then combine them using the symbols $\neg$, $\land$, and $\to$ to form the more complex statement we wanted to express. In a different context, of course, although logical operations like negation, conjunction, and implication will still be useful, the relevant "basic" statements might be entirely different—perhaps we want to reason not

[14] This "arrow" symbol stands for "implies".

[15] The brackets help us keep track of the order that we built the formula; more on this later.

[16] This "latch" symbol stands for "not".

[17] This "vee" symbol stands for "or".

about the weather but about the values of certain variables in a computer system, or about which cafes are open in the city, or whether certain numbers are prime, or the colours of our hats. For this reason we do not include specific statements like "it's raining" in our language, opting instead to keep the "basic" formulas of propositional logic entirely abstract, so that, intuitively, they can play any role we desire.

These "basic" formulas are called *primitive propositions*.[18] For convenience we let PROP denote the (countable) set of all primitive propositions,[19] and use letters like p, q, and r to denote them.[20] More complex formulas are built according to the following recursive clauses.[21]

- If φ is a formula, so is $\neg\varphi$ (negation).
- If φ and ψ are formulas, so is $\varphi \wedge \psi$ (conjunction).
- If φ and ψ are formulas, so is $\varphi \vee \psi$ (disjunction).
- If φ and ψ are formulas, so is $\varphi \rightarrow \psi$ (implication).

This way of defining a language recursively is often summarized succinctly using the following notation:[22]

$$\varphi ::= p \mid \neg\varphi \mid \varphi \wedge \psi \mid \varphi \vee \psi \mid \varphi \rightarrow \psi,$$

where $p \in$ PROP. Intuitively, this notation might be read from left to right as follows:

> What counts as a formula, φ? Well, it's either a primitive proposition $p \in$ PROP, or else it's the negation of another formula, or the conjunction of two formulas, or the disjunction of two formulas, or an implication from one formula to another.

Let $\mathcal{L}(\neg, \wedge, \vee, \rightarrow)$ denote the set of all formulas so defined; this is called the **language of propositional logic**. The symbols $\neg$, $\wedge$, $\vee$, and $\rightarrow$ are called *connectives*; the last three are called *binary* connectives because they connect two formulas, while negation is (somewhat oxymoronically) referred to as a *unary* connective.

[18] Or sometimes *atomic propositions*.

[19] If the concept of a set or the notion of "countability" is unfamiliar, see Appendices A.1 and A.4.

[20] This keeps the formalism fully general, avoiding commitment to any particular interpretation of the basic statements.

[21] Here and throughout this book, Greek letters like φ and ψ are used as variables to range over formulas in the given language. These are sometimes called *metavariables*, emphasizing the fact that φ and ψ are not themselves symbols in the propositional language, but rather notation that we are using to reason *about* the propositional language. Thus we distinguish between the *object language*—which we are about to define carefully—and the *metalanguage*—which we use to reason *about* the object language, and which is adopted by convention rather than formally defined.

[22] This presentation of the syntax is sometimes referred to as "Backus–Naur form".

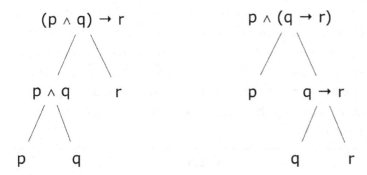

Figure 1.2 Parse trees for $(p \land q) \to r$ and $p \land (q \to r)$, respectively.

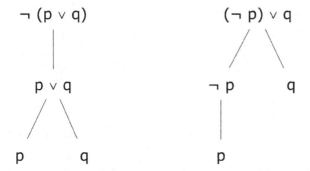

Figure 1.3 Parse trees for $\neg(p \lor q)$ and $(\neg p) \lor q$, the wide scope and narrow scope readings, respectively.

Each formula should be understood as having a *unique* step-by-step construction from the primitive propositions.[23] This means that in practice, when writing down formulas, we often need to make use of parentheses to disambiguate certain expressions. For example, above we considered the formula $(p \land q) \to r$. Notice the brackets serve to "remember" the fact that we started by combining p and q to form $p \land q$. Contrast this with a process where we first form the formula $q \to r$, and then from this create $p \land (q \to r)$. Because its construction process was different, this counts as a *different formula* from $(p \land q) \to r$; crucially, without the brackets, they would look exactly the same: $p \land q \to r$.

[23] This unique construction is called its *parse tree*. It is often visualized using a tree-like diagram as in Figures 1.1, 1.2, and 1.3.

[24] These are called "wide scope" and "narrow scope" readings, respectively. In general, the *scope* of a logical connective in a formula is the formula that it was applied to in the construction process; visually, this is the formula immediately below its lowest occurrence in the parse tree. So the scope of the negation in ¬($p \lor q$) is $p \lor q$, while its scope in (¬p) ∨ q is just p.

Similar reasoning applies when we consider ¬$p \lor q$, which could be interpreted either as ¬($p \lor q$) or (¬p) ∨ q.[24] In order to cut down on the number of parentheses in our formulas (and thereby improve readability), we adopt the (widespread) convention of interpreting negation as taking narrow scope in the absence of parentheses. Thus, when we write ¬$p \lor q$, we always mean (¬p) ∨ q.

Exercise 1

Draw parse trees for the following formulas.

(a) $p \to (q \to p)$
(b) $(p \to q) \to p$
(c) $(p \land q) \to (p \lor q)$
(d) $p \land ((q \to p) \lor q)$
(e) ¬¬$p \to p$
(f) ¬(¬$p \to p$)
(g) ¬¬($p \to p$)

Solution: page 194

The structure of the set $\mathcal{L}(\neg, \land, \lor, \to)$ makes it amenable to *recursive definitions*. The basic idea is to take advantage of the fact that the set is composed of "basic" elements (in this case, primitive propositions), with all other elements "built" from the basic elements by (repeatedly) applying certain operations. This means we can define a function on this set by first specifying the output on primitive propositions, and then extending the definition to handle the connectives.

For a concrete example, we might recursively define the *length* function on formulas as follows:[25]

[25] Here and elsewhere we will use ∗ to stand for an arbitrary binary connective, that is, any of ∧, ∨, or →.

$$length(p) = 1, \text{ for all } p \in \text{PROP}$$
$$length(\neg \varphi) = length(\varphi) + 1$$
$$length(\varphi * \psi) = length(\varphi) + length(\psi) + 1.$$

Intuitively: a single primitive proposition has length 1; adding a negation symbol to φ increases the length of this formula by 1;

and finally, joining two formulas φ and ψ with a binary connective increases the combined length of φ and ψ by 1. A key feature of a recursive definition is that the output on complex formulas is defined by reference to the output on simpler formulas, a process that continues all the way down to the primitive propositions—which are given explicitly. So, for instance, we can compute

$$length(\neg(p \rightarrow \neg q)) = length(p \rightarrow \neg q) + 1$$
$$= (length(p) + length(\neg q) + 1) + 1$$
$$= (1 + (length(q) + 1) + 1) + 1$$
$$= (1 + (1 + 1) + 1) + 1$$
$$= 5.$$

Here we see concretely that the *length* function essentially counts the total number of symbols in a given formula, interpreting each primitive proposition and each connective as a single symbol.[26]

[26] And ignoring parentheses.

For a slightly more sophisticated example, define *rank* recursively as follows:

$$rank(p) = 0, \text{ for all } p \in \text{PROP}$$
$$rank(\neg\varphi) = rank(\varphi) + 1$$
$$rank(\varphi * \psi) = \max\{rank(\varphi), rank(\psi)\} + 1.$$

With a little experimentation, you can convince yourself that $rank(\varphi)$ captures the depth of the parse tree of φ, that is, how many "layers" it has below the top.[27]

[27] So, for example, both of the parse trees pictured in Figures 1.2 and 1.3 have depth 2—and therefore the corresponding formulas ought to have rank 2. By contrast, the parse tree shown in Figure 1.1 has depth 3, so the corresponding formula ought to have rank 3. Try to verify this!

Exercise 2

Provide recursive definitions for the following.

(a) The function $conn(\varphi)$ returning the number of connectives in φ (counting duplicates—so for instance $conn(\neg\neg p) = 2$).

> (b) The function $sub(\varphi)$ returning the set of all formulas occurring in the parse tree for φ. These are called the *subformulas* of φ.
>
> Solution: page 195

How can we prove things about recursively defined objects, like sets of formulas? One powerful technique that takes advantage of the recursive nature of the set is *structural induction*. If you're already familiar with the concept of mathematical induction on the natural numbers, $\mathbb{N}$,[28] you can draw on that intuition.[29] The natural numbers themselves can be viewed as a recursively defined set: the "basic" element is 0, and the (sole) operation that builds more complex elements from those already constructed is "adding 1"; moreover, every natural number (and only natural numbers) are formed in this way, and the representation is unique.[30]

It turns out that the recursive structure of $\mathcal{L}(\neg, \wedge, \vee, \rightarrow)$ allows us to do induction in essentially the same way we do on $\mathbb{N}$. Here, the goal is to prove that some property holds true for every formula in $\mathcal{L}(\neg, \wedge, \vee, \rightarrow)$.

Let $P(\varphi)$ denote the claim that this property, whatever it is, is true for φ. Our "base case" is to prove that $P(p)$ holds for each primitive proposition $p \in$ PROP. And the "inductive step", as before, corresponds to the operations that build new elements from old: the connectives. This means that there are actually *four* inductive steps: prove that whenever $P(\varphi)$ is true, so is $P(\neg\varphi)$, and whenever $P(\varphi)$ and $P(\psi)$ are true, so are $P(\varphi \wedge \psi)$, $P(\varphi \vee \psi)$, and $P(\varphi \rightarrow \psi)$. An example will be helpful.

Proposition 1 *For every formula φ, length$(\varphi) >$ conn(φ).*[31]

Proof Induction on φ. For the base case, we observe that $length(p) = 1 > 0 = conn(p)$. Next, for the inductive steps, we suppose the result holds for φ and ψ, and try to show it also holds for $\neg\varphi$ and for $\varphi * \psi$. And indeed, we have:

[28] Recall that $\mathbb{N} = \{0, 1, 2, 3, \ldots\}$.

[29] See Appendix A.3 for an introduction or a refresher on this topic.

[30] That is, there's only one way of expressing any given natural number as $0 + 1 + 1 + \cdots + 1$.

[31] Of course, this requires the right definition of *conn*; see Exercise 2 and the solution thereto.

$$length(\neg\varphi) = length(\varphi) + 1 > conn(\varphi) + 1 = conn(\neg\varphi),$$

and

$$length(\varphi * \psi) = length(\varphi) + length(\psi) + 1$$
$$> conn(\varphi) + conn(\psi) + 1$$
$$= conn(\varphi * \psi). \qquad \square$$

Exercise 3

Is it true that for all $\varphi \in \mathcal{L}(\neg, \wedge, \vee, \rightarrow)$, $length(\varphi) > rank(\varphi)$? Justify your answer with an inductive proof or a counter-example.

Solution: page 195

Exercise 4

Is it true that for all $\varphi \in \mathcal{L}(\neg, \wedge, \vee, \rightarrow)$, $rank(\varphi) \geq conn(\varphi)$? Justify your answer with an inductive proof or a counter-example.

Exercise 5

For each $\varphi \in \mathcal{L}(\neg, \wedge, \vee, \rightarrow)$, let $f(\varphi)$ denote the number of atomic propositions occurring in φ (counting each instance separately, so for example $f(p \rightarrow p) = 2$), and let $g(\varphi)$ denote the number of times the symbol "$\wedge$" occurs in φ.

(a) Provide recursive definitions for f and g.
(b) Prove that for all φ, $f(\varphi) > g(\varphi)$.

Exercise 6

Show that a formula with n connectives has at most $2n + 1$ subformulas.

Solution: page 196

Exercise 7

Prove that $|sub(\varphi)| \leq 2^{rank(\varphi)+1} - 1$.

We will make *heavy* use of recursive definitions and structural induction in this book.

1.2 Semantics: valuations and truth tables

As noted, the language we have defined is just a bunch of symbols governed by rules telling us how we are allowed to string them together. While we might *refer* to a symbol like "$\wedge$" using the word "and", so far there is no actual connection linking this symbol to the specific concept of "andness", and likewise for the other connectives.

One way of establishing such a link is through the notion of *truth*. That is, we ask: when is a formula like $\varphi \wedge \psi$ true? The answer can be defined recursively in a way that encodes the meanings of the logical connectives.

As always for a recursive definition we need a base case, which here corresponds to a truth value for each primitive proposition p. Since primitive propositions are meant to be completely abstract, there is no particular reason to think that any given one is true or false, so we simply assign their truth values arbitrarily:[32] a **valuation** is a function $v \colon \mathrm{PROP} \rightarrow \{true, false\}$. This notion of truth is then extended recursively to a function[33]

$$[\![\cdot]\!]_v \colon \mathcal{L}(\neg, \wedge, \vee, \rightarrow) \rightarrow \{true, false\}$$

defined on all formulas as follows:[34]

$$[\![p]\!]_v = true \text{ iff } v(p) = true$$

$$[\![\neg\varphi]\!]_v = true \text{ iff } [\![\varphi]\!]_v = false$$

$$[\![\varphi \wedge \psi]\!]_v = true \text{ iff } [\![\varphi]\!]_v = true \text{ and } [\![\psi]\!]_v = true$$

$$[\![\varphi \vee \psi]\!]_v = true \text{ iff } [\![\varphi]\!]_v = true \text{ or } [\![\psi]\!]_v = true$$

$$[\![\varphi \rightarrow \psi]\!]_v = true \text{ iff } [\![\varphi]\!]_v = false \text{ or } [\![\psi]\!]_v = true.$$

[32] If this arbitrariness worries you, read on—we will see in the definition of a *tautology* why it is justified.

[33] Notation here may be a bit unfamiliar: normally the value of a function $f(\cdot)$ on an input x is written "$f(x)$". Here the function "symbol" is the double square brackets, and we denote its value on an input φ by $[\![\varphi]\!]_v$.

[34] The abbreviation "iff" stands for "if and only if"; in other words, it means the left-hand side holds *exactly* when the right-hand side does. So, for instance, the third line states not only that $\varphi \wedge \psi$ is true whenever both φ and ψ are, but *also* that $\varphi \wedge \psi$ is *not* true (so, false) whenever φ and ψ are not both true.

Here, given any formula φ, the expression "$[\![\varphi]\!]_v$" should be read as "the truth value of φ given the valuation v".[35] So the topmost line of the above definition specifies that the truth value of a primitive proposition is determined directly by v, while the remaining lines tell us how to extend this notion of truth to more complex formulas.

This definition of truth captures the "truth functional" essence of the logical connectives, namely, that the truth of a compound formula depends entirely on the truth of its component parts. Notice how the recursive definitions of conjunction and disjunction actually use the English words "and" and "or", respectively. Similarly, the definition of negation relies on the fact that there are only two truth values, so being false is the same as *not* being true.

Famously, the recursive clause for implication is not so intuitive.[36] For example, "If today is Saturday, then I'm in Buffalo" isn't a statement we tend to think of as being true simply because today isn't Saturday. It would be more natural to interpret it as saying something about where I *will* be on Saturday.[37] At root, this is because the concept of implication is very rich, and we cannot hope to fully capture it with such simple mathematical models.[38] Instead, we try to capture something that seems to line up with some sense of implication in at least some contexts.[39]

So how can we make sense of the recursive definition as given? Consider a statement like "If a number is prime and greater than 2, then it is odd". In evaluating such a claim, it seems natural to "ignore" cases where the antecedent is false.[40] After all, the number 14 is not a counterexample to this claim just because "14 is prime and greater than 2" is false. To falsify this implication, one would need an example of a number that makes the antedent true *and* makes the consequent false. This is what the recursive clause is trying to capture: that an implication counts as false only when the antecedent is true and the consequent is false; thus, an implication counts as true

[35] Of course, the formula φ might get assigned different truth values under different valuations.

[36] Literally hundreds of articles have been written on how to understand and mathematically model implication and various sorts of conditional statements.

[37] Or where I'm planning to be. Or where I *would* be on Saturday.

[38] To be fair, the same point might also be made for other connectives. For example, "I flipped the switch and the lights turned on" does not seem to mean quite the same thing as "The lights turned on and I flipped the switch", yet the recursive clause for $\wedge$ simply asks whether both conjuncts are true, with no care for their order.

[39] For this reason, the symbol "$\rightarrow$" in propositional logic is often called *material implication*, to emphasize the fact that it is not the "usual" notion of implication— whatever that is.

[40] In an implication $\varphi \rightarrow \psi$, we call φ the *antecedent* and ψ the *consequent*.

Table 1.1 The truth table for conjunction.

φ	$\wedge$	ψ
T	T	T
T	F	F
F	F	T
F	F	F

Table 1.2 Truth tables for disjunction, material implication, and negation.

φ	$\vee$	ψ	φ	$\rightarrow$	ψ		$\neg$	φ
T	T	T	T	T	T			
T	T	F	T	F	F		F	T
F	T	T	F	T	T		T	F
F	F	F	F	T	F			

if either the antecedent is false (in which case, intuitively, it doesn't apply) or the consequent is true.

The recursive definition of truth given above is often summarized by using *truth tables*; for example, the truth table for conjunction can be drawn as in Table 1.1. Here, of course, "T" stands for true and "F" for false. The way to read a truth table is to think of each row as capturing one possible combination of truth values for the formulas represented by variables (φ, ψ, etc.). Then, underneath the connectives, one writes the truth value corresponding to the formula that is built from that connective. So we see in Table 1.1 that $\varphi \wedge \psi$ is true only when φ and ψ are both true.

Truth tables for the other connectives are shown in Table 1.2.

Exercise 8

Provide truth tables for each of the following binary connectives.

(a) $\varphi \leftrightarrow \psi$, read "$\varphi$ if and only if ψ".
(b) $\varphi \oplus \psi$, read "either φ or ψ but not both" (aka "exclusive or").

(c) $\varphi | \psi$, read "not both φ and ψ" (aka "nand", the Sheffer stroke).

(d) $\varphi \downarrow \psi$, read "neither φ nor ψ" (aka "nor").

How many distinct truth tables are there for binary connectives?

Solution: page 196

Truth tables can also be used to display the recursive evaluation of truth for more complex formulas, as demonstrated in the proof of the following claim.

Proposition 2 *The formula* $\neg(\varphi \wedge \psi) \rightarrow (\neg\varphi \vee \neg\psi)$ *is true under all valuations.*

Proof We fill in the truth table for this formula step-by-step. Note that the formula contains two occurrences of φ and two occurrences of ψ, and since these are instances of the same formula, no matter what valuation we choose, they will always get the same truth value. Thus we begin with the following.

Table 1.3 Begin by filling in all possible combinations of truth values for the formulas φ and ψ.

¬	(φ	∧	ψ)	→	(¬	φ	∨	¬	ψ)
	T		T			T			T
	T		F			T			F
	F		T			F			T
	F		F			F			F

Next, use the truth tables for ∧ and ¬ to fill in the truth values that can be determined from the existing entries.

Table 1.4 Fill in the truth values determined by those of φ and ψ.

¬	(φ	∧	ψ)	→	(¬	φ	∨	¬	ψ)
	T	T	T		F	T		F	T
	T	F	F		F	T		T	F
	F	F	T		T	F		F	T
	F	F	F		T	F		T	F

Now iterate this process: use the truth tables for ¬ and ∨ to determine the truth values "one level up" in the parse tree.

Table 1.5 Iterate the process.

¬	(φ	∧	ψ)	→	(¬	φ	∨	¬	ψ)
F	T	T	T		F	T	F	F	T
T	T	F	F		F	T	T	T	F
T	F	F	T		T	F	T	F	T
T	F	F	F		T	F	T	T	F

Finally, using the truth table for →, complete the last column.

Table 1.6 The final, completed truth table.

¬	(φ	∧	ψ)	→	(¬	φ	∨	¬	ψ)
F	T	T	T	T	F	T	F	F	T
T	T	F	F	T	F	T	T	T	F
T	F	F	T	T	T	F	T	F	T
T	F	F	F	T	T	F	T	T	F

Since this final column of the truth table consists entirely of Ts, we have shown that the formula ¬(φ∧ψ) → (¬φ∨¬ψ) is true under all valuations. □

Exercise 9

Fill in truth tables for the following formulas.

(a) (φ ∨ ψ) → φ
(b) (φ → ψ) ∧ (ψ → φ)
(c) φ ∧ φ
(d) ¬(φ ∧ ¬ψ)
(e) (φ ∨ ψ) → (φ ∧ ψ)
(f) (φ ∧ ψ) ∨ χ
(g) φ ∧ (ψ ∨ χ)

Solution: page 197

This definition of truth might feel a bit unsatisfying because it all bottoms out in the aforementioned arbitrary assignment of truth to the primitive propositions. How does this arbitrary assignment jive with the supposedly "universal" character of logical truths? The answer comes in the following definition.

Say that φ is a **tautology**, and write $\vDash \varphi$, if for *all* valuations v, $[\![\varphi]\!]_v = true$. So tautologies are formulas that are "true in virtue of form"—that is, their truth does not depend on any contingent assignment of truth values to the primitive propositions, but is guaranteed by the form of the statement itself. For example, every formula of the form $\neg(\varphi \wedge \psi) \to (\neg\varphi \vee \neg\psi)$ is a tautology, as we saw in Proposition 2.[41]

It's not hard to see that there are many tautologies.[42]

Exercise 10

Show that every instance of each of the following formula schemes is a tautology.

(a) $\varphi \vee \neg\varphi$

(b) $\varphi \to \varphi$

(c) $(\varphi \wedge \psi) \leftrightarrow (\psi \wedge \varphi)$

(d) $\neg(\varphi \wedge \psi) \leftrightarrow (\neg\varphi \vee \neg\psi)$

(e) $(\varphi \to \psi) \leftrightarrow (\neg(\varphi \wedge \neg\psi))$

(f) $\varphi \to (\psi \to \varphi)$

(g) $(\varphi \to (\psi \to \chi)) \to ((\varphi \to \psi) \to (\varphi \to \chi))$

(h) $(\neg\psi \to \neg\varphi) \to (\varphi \to \psi)$

Solution: page 197

---

[41] An important subtlety here is that $\neg(\varphi \wedge \psi) \to (\neg\varphi \vee \neg\psi)$ is not a single formula but rather a family of formulas, since φ and ψ are variables that themselves range over formulas. It is therefore referred to as a *formula scheme*, and the members of this family as *instances* of the scheme. Table 1.6 shows that every instance of this formula scheme is a tautology.

[42] Here and henceforth, we write $\varphi \leftrightarrow \psi$ as an abbreviation for

$$(\varphi \to \psi) \wedge (\psi \to \varphi).$$

This connective is sometimes called *bi-implication* and read "if and only if". Compare Exercises 8(a) and 9(b).

1.3 Logical equivalence and adequacy

In this section we take a deeper dive into the semantics of propositional logic we have just defined, exploring the notion of logical equivalence, expressivity of different languages, and (briefly) the concept of "logical space". Fair warning: the material here is a notch higher in terms of difficulty and abstraction; this section can be safely skimmed or skipped on a first read-through.

Two formulas φ and ψ are called *logically equivalent* if $\models \varphi \leftrightarrow \psi$.[43] To capture this succinctly we use the notation $\varphi \approx \psi$. In fact, this relation $\approx$ on formulas is an *equivalence relation*; recall this means it is *reflexive*,[44] *symmetric*,[45] and *transitive*.[46]

[43] It is easy to check that this means every valuation gives the same truth value to φ as it does to ψ. Again, see Exercises 8(a) and 9(b).

[44] $\forall \varphi (\varphi \approx \varphi)$

[45] $\forall \varphi, \psi (\varphi \approx \psi \Rightarrow \psi \approx \varphi)$

[46] $\forall \varphi, \psi, \chi ((\varphi \approx \psi \ \& \ \psi \approx \chi) \Rightarrow \varphi \approx \chi)$

Exercise 11

Prove that $\approx$ is an equivalence relation on $\mathcal{L}(\neg, \wedge, \vee, \rightarrow)$.

Exercise 12

Let $\varphi[\psi/p]$ denote the **substitution of ψ for p in φ**, which is recursively defined (on φ) as follows:

$$q[\psi/p] = \begin{cases} \psi & \text{if } q = p \\ q & \text{otherwise} \end{cases}$$

$$(\neg \varphi)[\psi/p] = \neg(\varphi[\psi/p])$$

$$(\varphi_1 * \varphi_2)[\psi/p] = (\varphi_1[\psi/p]) * (\varphi_2[\psi/p]).$$

(a) Show that $[\![\psi_1]\!]_v = [\![\psi_2]\!]_v$ implies $[\![\varphi[\psi_1/p]]\!]_v = [\![\varphi[\psi_2/p]]\!]_v$.

(b) Conclude from part (a) that the *Substitution Theorem* holds:

$$\models \psi_1 \leftrightarrow \psi_2 \text{ implies } \models \varphi[\psi_1/p] \leftrightarrow \varphi[\psi_2/p].$$

Solution: page 198

It is possible to show that every formula φ is logically equivalent to a formula $\varphi' \in \mathcal{L}(\neg, \wedge)$.[47] In other words, up to logical equivalence, we can express everything in the propositional language using just the connectives $\neg$ and $\wedge$. Sets of connectives with this property are called *adequate*.

[47] Where, of course, this is the language recursively defined by

$$\varphi ::= p \mid \neg \varphi \mid \varphi \wedge \psi.$$

How can we prove a claim like this? It's easy enough to see that a single occurrence of a connective like $\rightarrow$ can be "removed" from a formula using an equivalence like the one proved in Exercise 10(e). But what about a formula with multiple occurences of $\rightarrow$ (not to mention $\vee$)? We need to somehow iterate the process. This is a job for structural induction.

Theorem 3 *For every formula $\varphi \in \mathcal{L}(\neg, \wedge, \vee, \rightarrow)$, there is a formula $\varphi' \in \mathcal{L}(\neg, \wedge)$ such that $\varphi' \approx \varphi$.*

Proof Induction on the structure of φ. The base case, when φ is a primitive proposition, is trivial because every primitive proposition in $\mathcal{L}(\neg, \wedge, \vee, \rightarrow)$ is also in $\mathcal{L}(\neg, \wedge)$.

Suppose inductively the result holds for φ, and let φ' be a formula in $\mathcal{L}(\neg, \wedge)$ with $\varphi' \approx \varphi$. Then it is easy to check that $\neg\varphi' \approx \neg\varphi$, and so since $\neg\varphi' \in \mathcal{L}(\neg, \wedge)$ (since φ' is), this completes the inductive step for $\neg$.

Next suppose inductively the result holds for φ and ψ, and let φ' and ψ' be formulas in $\mathcal{L}(\neg, \wedge)$ that are logically equivalent to φ and ψ, respectively. Clearly then $\varphi' \wedge \psi' \approx \varphi \wedge \psi$, and so since $\varphi' \wedge \psi' \in \mathcal{L}(\neg, \wedge)$, this completes the inductive step for $\wedge$.

These were the easy cases, dealing with connectives that are allowed in both languages. Now consider the case of $\rightarrow$. Following Exercise 10(e), we show that $\neg(\varphi' \wedge \neg\psi')$ is logically equivalent to $\varphi \rightarrow \psi$; indeed, we have the following.[48]

(φ	$\rightarrow$	ψ)	$\leftrightarrow$	($\neg$	(φ'	$\wedge$	$\neg$	ψ'))
T	T	T	T	T	T	F	F	T
T	F	F	T	F	T	T	T	F
F	T	T	T	T	F	F	F	T
F	T	F	T	T	F	F	T	F

[48] Note that we only include rows where the truth values of φ and φ' coincide, and similarly for ψ and ψ', since we know these formulas are logically equivalent.

Since $\neg(\varphi' \wedge \neg\psi') \in \mathcal{L}(\neg, \wedge)$, this completes the inductive step for $\rightarrow$.

Finally, we consider the inductive step for $\vee$. Inspired by Exercise 10(d), we show that $\neg(\neg\varphi' \wedge \neg\psi')$ is logically equivalent to $\varphi \vee \psi$; indeed, we have the following.

(φ	$\vee$	ψ)	$\leftrightarrow$	($\neg$	($\neg$	φ'	$\wedge$	$\neg$	ψ'))
T	T	T	T	T	F	T	F	F	T
T	T	F	T	T	F	T	F	T	F
F	T	T	T	T	T	F	F	F	T
F	F	F	T	F	T	F	F	T	F

As previously, since $\neg(\neg\varphi' \wedge \neg\psi') \in \mathcal{L}(\neg, \wedge)$, this completes the inductive step for $\vee$. $\qquad\square$

Exercise 13

Show that $\{\neg, \vee\}$, $\{\neg, \rightarrow\}$, $\{\,|\,\}$, and $\{\downarrow\}$ are also adequate sets of connectives.

Solution: page 199

Exercise 14

Show that for each $\varphi \in \mathcal{L}(\vee, \rightarrow)$ there is a valuation v such that $[\![\varphi]\!]_v = true$, and use this to conclude that $\{\vee, \rightarrow\}$ is *not* an adequate set of connectives.

Solution: page 199

Exercise 15

Show that $\{\neg, \leftrightarrow\}$ is not an adequate set of connectives.

Hint: page 200

Exercise 16

Show that an adequate set of connectives can express any truth-functional binary connective: more precisely, show that if $\triangle$ is a binary connective interpreted according to any truth table, and A is an adequate set of connectives, then for each $\varphi \in \mathcal{L}(A \cup \{\triangle\})$ there is a logically equivalent $\varphi' \in \mathcal{L}(A)$.

This notion of adequacy is fundamentally semantic, since it relies on the definition of logical equivalence. So we have shown that—as long as we keep the semantics unchanged—we *could* have just started with the language $\mathcal{L}(\neg, \wedge)$ from the outset, and not bothered with disjunction or implication, except perhaps to introduce them as abbreviations for convenience, as we did with bi-implication. This observation will be useful for us going forward.

We conclude this section by introducing an abstract but quite useful perspective on logical formulas.[49] The notion of logical equivalence, being an equivalence relation, induces a partition on the set of all formulas.[50] Intuitively, by squishing together formulas that are logically equivalent, we can produce something like "points in logical space" that serve to capture a sort of "pure" notion of "logical strength".

Of course, this is very vague. To make it precise, consider the equivalence classes induced by logical equivalence:[51]

$$[\varphi] := \{\psi \; : \; \psi \approx \varphi\}.$$

Each such class represents a different point in what we referred to above as "logical space": at one extreme is the class $[\top]$,[52] collecting all the tautologies; at the other extreme is $[\bot]$,[53] the class of all *contradictions*; in between are all the classes corresponding to formulas that are neither tautologies nor contradictions, grouped together according to logical equivalence. For example, $p \wedge q$ and $q \wedge p$, although distinct as formulas, are logically equivalent and thus correspond to the same "point in logical space"—that is, the same equivalence class. Similarly, the formulas p, $p \wedge p$, $p \vee p$, and $\top \rightarrow p$ are all logically equivalent and so again they are grouped together into one single "point" in logical space—one single equivalence class. Moreover, we can define an order on these equivalence classes by setting $[\varphi] \leq [\psi]$ iff $\vDash \varphi \rightarrow \psi$, in which case $[\top]$ is at the "top" and $[\bot]$ is at the "bottom", giving at least some credence to the "space" metaphor.

---

[49] These paragraphs can safely be skipped or skimmed on a first or even a second read-through.

[50] For a review of (or introduction to) equivalence relations and partitions, see Appendix A.2.

[51] To streamline the notation here, we omit the subscript $\approx$.

[52] Where $\top$ is an abbreviation for $p \vee \neg p$.

[53] Where $\bot$ abbreviates $\neg \top$.

Exercise 17

Consider the set C consisting of all valuation functions for propositional logic (based on some fixed set of primitive propositions $\{p_1, p_2, \ldots\}$). For each formula φ, define

$$\llbracket \varphi \rrbracket = \{v \; : \; \llbracket \varphi \rrbracket_v = true\}.$$

Show that:

(a) $\llbracket \varphi \wedge \psi \rrbracket = \llbracket \varphi \rrbracket \cap \llbracket \psi \rrbracket$

(b) $[\![\varphi \vee \psi]\!] = [\![\varphi]\!] \cup [\![\psi]\!]$

(c) $[\![\neg\varphi]\!] = [\![\varphi]\!]^c$ (the complement)

(d) $[\![\top]\!] = C$ and $[\![\bot]\!] = \varnothing$

(e) $\vDash \varphi \rightarrow \psi$ if and only if $[\![\varphi]\!] \subseteq [\![\psi]\!]$.

What is the relationship between $\{[\![\varphi]\!] : \varphi \in \mathcal{L}(\neg, \wedge, \vee, \rightarrow)\}$ and $\mathcal{L}(\neg, \wedge, \vee, \rightarrow)/\approx$? (And why did I name the set above C?)

Solution: page 200

1.4 Deduction

There is another way of linking the intended meaning of "and" to the $\wedge$ symbol: through the notion of deduction. In the previous section we focussed on the idea of *truth*, and gave meaning to the logical symbols by defining what would make formulas containing them true. Here, we (temporarily) eschew the concept of truth and instead focus on *reasoning*. By specifying what we could *deduce* from a formula like $\varphi \wedge \psi$, we are able to capture part of the intended meaning of the symbol $\wedge$ without reference to truth values.

For example, we might declare that from $\varphi \wedge \psi$ we can deduce both φ and ψ, which we might encode using formula schemes: $(\varphi \wedge \psi) \rightarrow \varphi$ and $(\varphi \wedge \psi) \rightarrow \psi$. Such declarations are called "axioms"; intuitively, they are formulas that we treat as foundational assumptions, requiring no justification. In a similar vein, we might consider taking as axioms formula schemes like $\varphi \rightarrow (\varphi \vee \psi)$, or $\neg\neg\varphi \rightarrow \varphi$, or even something a bit fancier like $(\varphi \rightarrow \psi) \rightarrow (\neg\psi \rightarrow \neg\varphi)$.

All this is still informal—we are laying the intuition here for a formal definition of *deduction* in propositional logic. Recall the view of logic as "the mathematics of mathematics": in that sense, what we are really doing here is beginning to formalize the mathematical notion of a *proof*.[54] To do so, we must answer two questions.

- What are the foundational assumptions that we can appeal to without justification?

[54] Of course, there is a tight connection between provability and truth; this is the whole point of the next section (Section 1.5) on *soundness and completeness*.

- How can we infer new things from what we have already deduced/proven?[55]

The answer to the first question comes, as we began to explore above, by specifying a set of *axioms*: a collection of formulas that we simply declare to be true by fiat.[56] There are many candidate axioms we might consider, but for concreteness we will focus on the collection given by all instances of the following three schemes:[57]

(L1) $\varphi \to (\psi \to \varphi)$

(L2) $(\varphi \to (\psi \to \chi)) \to ((\varphi \to \psi) \to (\varphi \to \chi))$

(L3) $(\neg\psi \to \neg\varphi) \to (\varphi \to \psi)$.

All of these axioms are written in the language $\mathcal{L}(\neg, \to)$. Henceforth, we will work with this smaller (but no less expressive!) language, as justified by Exercise 13.[58] This will simplify future proofs, and it allows us to avoid having to write down extra axioms specifically governing the use of the other connectives. It is worth pointing out that this sort of "minimalism" will actually make it *harder* to come up with formal deductions (as defined below), but it will streamline our proofs *about* deductions.[59] Since our primary goal is the latter, this is a reasonable tradeoff!

The answer to the second question is given by specifying one or more *rules of inference*. In fact, we can get away with only one: modus ponens. This rule essentially tells us that from φ and $\varphi \to \psi$ we are licensed to infer ψ.

Axioms and rules of inference come together in the process of reasoning—if axioms are the foundations of reasoning, rules of inference are the engine that drives it forward, allowing us to move beyond the axioms. We want to define a *deduction* to mirror the structure of mathematical reasoning, that is, proof—intuitively, it consists in a finite series of "lines",[60] each of which is somehow "justified", either because it's an axiom,[61] or because it can be inferred from *previous* lines using a rule of inference.

Formally, a **deduction** is a finite sequence of formulas $\varphi_1, \ldots, \varphi_n$ in $\mathcal{L}(\neg, \to)$ such that for each $1 \le i \le n$, either:

[55] "Deduction" and "proof" are sometimes treated as synonyms; in this book, we will aim to use the former for the formal notions we define explicitly, and reserve the latter for the more colloquial, conventional notion that mathematicians take for granted.

[56] It's also important that these axioms be *effectively presented*, that is, that we can (algorithmically) recognize which formulas are axioms—intuitively, it's hard to write a proof if you can't recognize which things are axioms!

[57] Note that each of these is a tautology—see Exercise 10. This will be important for proving soundness.

[58] Thus, as we have done previously, we are essentially treating formulas involving the other connectives as abbreviations for logically equivalent formulas in the language $\mathcal{L}(\neg, \to)$.

[59] Said differently: it will make it harder to prove things in the *object language*, but easier to reason in the *metalanguage*.

[60] Like the lines of a proof!

[61] And therefore, in the sense described above, requires no further justification.

[62] That is, an instance of (L1), (L2), or (L3).

(1) φ_i is an axiom,[62] or

(2) there exist $j, k < i$ such that $\varphi_k = \varphi_j \to \varphi_i$.

Such a sequence is said to be a *deduction of* φ_n; we write $\vdash \varphi$ to indicate that there exists a deduction of φ.

[63] Abbreviated by "MP".

Note that our rule of inference, modus ponens,[63] is implicit in this definition: the "line" φ_i can be justified by two previous "lines" of the form φ_j and $\varphi_j \to \varphi_i$.[64] This definition of deduction is thus in accordance with the intuitions above about how to write a mathematical proof: it says you can write down instances of axiom schemes whenever you want, and otherwise the only thing you can do is write lines that follow from previous lines by modus ponens.

[64] Since the lines are numbered, the insistence on applying modus ponens only to *previous* lines is expressed by the requirement that $j, k < i$.

Note also that we explicitly require deductions to be finite. This accords both with the commonsense notion of a proof—namely, that it has to end at some point—but also with a view of deduction as something *algorithmic*: a kind of finitistic pattern-matching that a computer or mechanical device could, in principle, duplicate.

Finally, it's crucial to observe that the "single turnstile" symbol ($\vdash$) we have introduced is *not* the same as the "double turnstile" symbol ($\vDash$) we used previously. Not only are they different symbols, but they mean *completely* different things. We write $\vDash \varphi$ to indicate that all valuation functions, when recursively extended to formulas by truth tables, make φ true. By contrast, we write $\vdash \varphi$ to mean that there is some finite sequence of formulas ending in φ that satisfies the conditions for being a deduction specified above. So $\vDash \varphi$ is a fundamentally semantic notion, having to do with truth, whereas $\vdash \varphi$ is completely proof-theoretic.[65]

[65] The utterly different ways these notions are defined is what makes it impressive that we can prove they are equivalent!

Let's consider an example.

Proposition 4 $\vdash p \to p$.

Proof We need to show there exists a deduction of $p \to p$, so it's enough to provide one.

$$\varphi_1 \ = \ p \to ((p \to p) \to p) \tag{L1}$$

$$\varphi_2 \ = \ (p \to ((p \to p) \to p)) \to ((p \to (p \to p)) \to (p \to p)) \tag{L2}$$

$$\varphi_3 \ = \ (p \to (p \to p)) \to (p \to p) \tag{1,2 MP}$$

$$\varphi_4 \ = \ p \to (p \to p) \tag{L1}$$

$$\varphi_5 \ = \ p \to p \tag{3,4 MP} \quad \square$$

Although this deduction is only five lines long, it's also fair to say that it's not very intuitive—especially given that the formula we wanted to prove was so simple. Why did we start with φ_1? Here's a bit of intuition: a decent guess is to start with $p \to (p \to p)$, from (L1). Then what is there to do? Well, we could use (L2) and (MP) to get $(p \to p) \to (p \to p)$. This is sort of close to what we want, but not quite it. It would be better if the antecedent was itself an instance of an axiom scheme, like perhaps $p \to (p \to p)$... working backwards, this thought motivates choosing the first line to be φ_1.

The simple fact is: formal deductions like these are *not easy* to come up with.[66] They do not tend to match up with "intuitive" proofs, partly because we are being so strict about what axioms we can use, and even stricter about what rules of inferences are allowed (only one!). The reason for this minimalism, as noted above, is simple: our goal here is *not* to become really good at writing formal deductions. The real goal is to articulate a formal notion of deduction that we can prove things *about*, and the more complicated that notion is, the harder it generally is to prove stuff about it.[67]

It's worth pointing out that nothing about the previous proof depended on the primitive proposition p—it could have been replaced with any formula φ, since all our axioms are given as schemes. Thus, we have the following.

Proposition 5 $\vdash \varphi \to \varphi.$

Proof As in the proof of Proposition 4, replacing each p with φ. $\quad \square$

[66] In some ways they seem more suitable for computers to work with. As alluded to above, this is part of the point of a deductive system—it's supposed to be "mechanical", in the sense that all of the individual steps are easily encoded algorithmically and could be performed by a sufficiently complex machine.

[67] There are many other ways of defining a notion of deduction, such as *sequent calculus* and so-called *natural deduction*, which you might (after some practice) find more intuitive to use. But, again, our main goal here is to prove meta-logical results *about* proofs, not to spend our time actually writing down proofs of propositional tautologies.

One way to begin to make deductions feel a bit more natural is by incorporating the idea of *assumptions*. Given a set of formulas $\Gamma \subseteq \mathcal{L}(\neg, \rightarrow)$, we can extend the definition of a deduction by simply allowing any formula in Γ to be written at any point in the deduction, similar to instances of axioms.[68] Formally, a **deduction from** Γ is a finite sequence of formulas $\varphi_1, \ldots, \varphi_n \in \mathcal{L}(\neg, \rightarrow)$ such that for each $1 \leq i \leq n$, either:

(1) φ_i is an axiom,

(2) $\varphi_i \in \Gamma$, or

(3) there exist $j, k < i$ such that $\varphi_k = \varphi_j \rightarrow \varphi_i$.

We write $\Gamma \vdash \varphi$ to indicate that there exists a deduction of φ from Γ.[69] Keep in mind that formulas in Γ don't need to be tautologies—intuitively, they are simply assumptions we are momentarily taking on board to see what we can deduce from them.

[68] Unlike the axioms, however, formulas in Γ are not meant to be interpreted as schemes, but as individual formulas.

[69] When $\Gamma = \varnothing$ we recover the original definition of a deduction.

Proposition 6 $\{\varphi \rightarrow \psi, \psi \rightarrow \chi\} \vdash \varphi \rightarrow \chi$.

Proof As before, we provide a formal deduction—only this time, in addition to instances of axioms schemes, we can write the two formulas $\varphi \rightarrow \psi$ and $\psi \rightarrow \chi$ without justification at any point in the deduction.

$$
\begin{array}{llll}
\varphi_1 & = & \varphi \rightarrow \psi & \text{(assump.)} \\
\varphi_2 & = & \psi \rightarrow \chi & \text{(assump.)} \\
\varphi_3 & = & (\psi \rightarrow \chi) \rightarrow (\varphi \rightarrow (\psi \rightarrow \chi)) & \text{(L1)} \\
\varphi_4 & = & \varphi \rightarrow (\psi \rightarrow \chi) & \text{(2,3 MP)} \\
\varphi_5 & = & (\varphi \rightarrow (\psi \rightarrow \chi)) \rightarrow ((\varphi \rightarrow \psi) \rightarrow (\varphi \rightarrow \chi)) & \text{(L2)} \\
\varphi_6 & = & (\varphi \rightarrow \psi) \rightarrow (\varphi \rightarrow \chi) & \text{(4,5 MP)} \\
\varphi_7 & = & \varphi \rightarrow \chi & \text{(1,6 MP)} \square
\end{array}
$$

Exercise 18

Prove the following by writing out completely formal deductions.

(a) $\vdash (\varphi \rightarrow \psi) \rightarrow (\varphi \rightarrow \varphi)$

(b) $\vdash \varphi \rightarrow (\psi \rightarrow (\varphi \rightarrow \psi))$

(c) $\{\neg\varphi \rightarrow \neg\psi, \psi\} \vdash \varphi$

(d) $\{\neg\varphi\} \vdash \varphi \rightarrow \psi$

(e) $\vdash \neg\varphi \rightarrow (\varphi \rightarrow \psi)$

(f) $\{\varphi \rightarrow \psi, \psi \rightarrow \chi, \varphi\} \vdash \chi$

(g) $\{\varphi \rightarrow \psi, \neg(\psi \rightarrow \chi) \rightarrow \neg\varphi\} \vdash \varphi \rightarrow \chi$

Solution: page 201

In fact, these "temporary assumptions" can be made "permanent" by importing them as antecedents via material implication. For instance, in Exercise 18, we can prove that part (e) follows from part (d). In other words, if we can deduce that $\varphi \rightarrow \psi$ from the assumption that $\neg\varphi$, then we can also deduce that $\neg\varphi \rightarrow (\varphi \rightarrow \psi)$ from *no* assumptions.[70] This idea applies to all formulas, and is captured by the Deduction Theorem, which tells us how we can introduce and discharge assumptions. First we need an easy lemma.

Lemma 7 *If $\varphi_1, \ldots, \varphi_n$ is a deduction from Γ, then for all $1 \leq m \leq n$, the truncated sequence $\varphi_1, \ldots, \varphi_m$ is also a deduction from Γ.*

Proof All we must show is that $\varphi_1, \ldots, \varphi_m$ satisfies the definition of a deduction from Γ: namely, that each formula in this sequence is either an instance of an axiom scheme, a member of Γ, or else follows from two previous lines by modus ponens. Since $\varphi_1, \ldots, \varphi_n$ is a deduction from Γ, this follows immediately.[71] □

Theorem 8 (The Deduction Theorem) *For every set of formulas $\Gamma \subseteq \mathcal{L}(\neg, \rightarrow)$ and any formulas $\varphi, \psi \in \mathcal{L}(\neg, \rightarrow)$, we have*

$$\Gamma \cup \{\varphi\} \vdash \psi \text{ if and only if } \Gamma \vdash \varphi \rightarrow \psi.$$

Proof The "if" direction is fairly easy: if $\varphi_1, \ldots, \varphi_n$ is a deduction of $\varphi \rightarrow \psi$ from Γ, then the following is a deduction of ψ from $\Gamma \cup \{\varphi\}$.

[70] The converse is true as well, though it is less impressive.

[71] The key intuition here is that each "line" of a proof is justified solely by reference to *previous* lines: modus ponens can only look backwards.

$$\varphi_1$$
$$\varphi_2$$
$$\vdots$$

$$\varphi_n \quad = \quad \varphi \to \psi$$
$$\varphi_{n+1} \quad = \quad \varphi \qquad \text{(assump.)}$$
$$\varphi_{n+2} \quad = \quad \psi \qquad (n, (n+1)\ \text{MP})$$

The "only if" direction is somewhat harder: we will use induction on the *length of the deduction* of ψ. More precisely, given Γ and φ we will prove the following by induction on n: for all $n \geq 1$, for all ψ, if $\psi_1, \ldots, \psi_n$ is a deduction of ψ from $\Gamma \cup \{\varphi\}$, then $\Gamma \vdash \varphi \to \psi$.

First the base case, $n = 1$. In this case $\psi_1 = \psi$, and therefore ψ is either an instance of an axiom scheme, or else an assumption in $\Gamma \cup \{\varphi\}$. If ψ is an instance of an axiom scheme, then we can see that the following is a deduction of $\varphi \to \psi$ from Γ.

$$\varphi_1 \quad = \quad \psi \to (\varphi \to \psi) \quad \text{(L1)}$$
$$\varphi_2 \quad = \quad \psi \qquad\qquad\quad \text{(L1), (L2), or (L3)}$$
$$\varphi_3 \quad = \quad \varphi \to \psi \qquad\quad (1,2\ \text{MP})$$

If $\psi \in \Gamma$, then similarly we see that the following is a deduction of $\varphi \to \psi$ from Γ.

$$\varphi_1 \quad = \quad \psi \to (\varphi \to \psi) \quad \text{(L1)}$$
$$\varphi_2 \quad = \quad \psi \qquad\qquad\quad \text{(assump.)}$$
$$\varphi_3 \quad = \quad \varphi \to \psi \qquad\quad (1,2\ \text{MP})$$

Finally, if $\psi = \varphi$, then we must show that $\varphi \to \varphi$ can be deduced from Γ. But we've already proved (Proposition 5) the stronger fact that $\vdash \varphi \to \varphi$. This completes the base case.

Now suppose inductively that the result holds for all $k \leq n$; we will show it also holds for $n + 1$.[72] So let $\psi_1, \ldots, \psi_{n+1}$ be a deduction of ψ from $\Gamma \cup \{\varphi\}$; we must show that $\Gamma \vdash \varphi \to \psi$. If ψ is an instance of an axiom scheme or a member of $\Gamma \cup \{\varphi\}$, then the same reasoning as in the base case holds and we are done. So the only remaining possibility is that ψ follows from two previous lines of the deduction by modus ponens: that is, there exist $i, j < n + 1$ such that $\psi_i = \psi_j \to \psi$.

[72] Here we are using a version of induction called "strong induction": instead of assuming the result is true for n, we are assuming it is true for all numbers *up to and including n*. It's not hard to see that strong induction is equivalent to regular induction; see Appendix A.3.

This is where the inductive hypothesis comes in: both $\psi_1, \ldots, \psi_j$ and $\psi_1, \ldots, \psi_i$ are deductions from $\Gamma \cup \{\varphi\}$ of length at most n;[73] therefore, by the inductive hypothesis, we know that $\Gamma \vdash \varphi \rightarrow \psi_j$ and also $\Gamma \vdash \varphi \rightarrow (\psi_j \rightarrow \psi)$. We can then concatenate a deduction of $\varphi \rightarrow \psi_j$ from Γ with a deduction of $\varphi \rightarrow (\psi_j \rightarrow \psi)$ from Γ, and add a few final lines to produce a deduction of $\varphi \rightarrow \psi$ from Γ.

[73] This follows from Lemma 7: any truncation of a deduction from Γ is itself a deduction from Γ.

$$\varphi_1$$
$$\vdots$$

$$\varphi_m \quad = \quad \varphi \rightarrow \psi_j$$
$$\varphi_{m+1}$$
$$\vdots$$

$$\varphi_{m'} \quad = \quad \varphi \rightarrow (\psi_j \rightarrow \psi)$$
$$\varphi_{m'+1} \quad = \quad (\varphi \rightarrow (\psi_j \rightarrow \psi)) \rightarrow ((\varphi \rightarrow \psi_j) \qquad \text{(L2)}$$
$$\qquad \qquad \rightarrow (\varphi \rightarrow \psi))$$
$$\varphi_{m'+2} \quad = \quad (\varphi \rightarrow \psi_j) \rightarrow (\varphi \rightarrow \psi) \qquad ((m'),(m'+1)\text{ MP})$$
$$\varphi_{m'+3} \quad = \quad \varphi \rightarrow \psi \qquad\qquad\qquad (m,(m'+2)\text{ MP}) \quad \square$$

The Deduction Theorem is a powerful and useful tool. It tells us that to deduce an implication of the form $\varphi \rightarrow \psi$, we can start by assuming the antecedent φ, then deduce the consequent ψ from that assumption, and finally apply the Deduction Theorem to conclude that $\vdash \varphi \rightarrow \psi$. For example, applying the Deduction Theorem to Proposition 6, we obtain $\{\varphi \rightarrow \psi\} \vdash (\psi \rightarrow \chi) \rightarrow (\varphi \rightarrow \chi)$; applying it a second time yields $\vdash (\varphi \rightarrow \psi) \rightarrow ((\psi \rightarrow \chi) \rightarrow (\varphi \rightarrow \chi))$. Try proving this directly!

In fact, the idea behind this formula—that you can "chain together" implications—is useful enough that it's worthwhile to show that it can be treated as a rule of inference without changing what can be deduced.

Exercise 19

Define a **deduction+** of φ from Γ to be a finite sequence of formulas $\varphi_1, \ldots, \varphi_n$ such that $\varphi_n = \varphi$ and, for each $i \leq n$, either (1) φ_i is an instance of an axiom scheme, or (2) $\varphi_i \in \Gamma$, or (3) there are $j, k < i$ such that $\varphi_k = \varphi_j \to \varphi_i$, or (4) there are $j, k < i$ such that $\varphi_j = \psi_1 \to \psi_2$, $\varphi_k = \psi_2 \to \psi_3$, and $\varphi_i = \psi_1 \to \psi_3$.

Prove that there is a deduction+ of φ from Γ if and only if there is a deduction of φ from Γ.

Solution: page 201

In light of Exercise 19, we can freely apply this new rule of inference in any deduction, safe in the knowledge that we *could* have done without it. It is sometimes called *hypothetical syllogism*.[74]

[74] Abbreviated by "HS".

We close this section with one final example of a deduction— partly because it is a very useful formula to know is deducible, and partly to convince you that even with the Deduction Theorem (and hypothetical syllogism as an additional rule of inference) in hand, deductions are still *hard*.[75]

[75] If you're itching to try your hand at a really challenging deduction, try this one before you look at the proof.

Proposition 9 $\vdash (\neg\varphi \to \varphi) \to \varphi$.

Proof By the Deduction Theorem, it is enough to show that $\{\neg\varphi \to \varphi\} \vdash \varphi$.

φ_1	$= \neg\varphi \to (\neg\neg(\neg\varphi \to \varphi) \to \neg\varphi)$	(L1)
φ_2	$= (\neg\neg(\neg\varphi \to \varphi) \to \neg\varphi) \to (\varphi \to \neg(\neg\varphi \to \varphi))$	(L3)
φ_3	$= \neg\varphi \to (\varphi \to \neg(\neg\varphi \to \varphi))$	(1,2 HS)
φ_4	$= (\neg\varphi \to (\varphi \to \neg(\neg\varphi \to \varphi)))$	(L2)
	$\quad \to ((\neg\varphi \to \varphi) \to (\neg\varphi \to \neg(\neg\varphi \to \varphi)))$	
φ_5	$= (\neg\varphi \to \varphi) \to (\neg\varphi \to \neg(\neg\varphi \to \varphi))$	(3,4 MP)
φ_6	$= \neg\varphi \to \varphi$	(assump.)
φ_7	$= \neg\varphi \to \neg(\neg\varphi \to \varphi)$	(5,6 MP)
φ_8	$= (\neg\varphi \to \neg(\neg\varphi \to \varphi)) \to ((\neg\varphi \to \varphi) \to \varphi)$	L3
φ_9	$= (\neg\varphi \to \varphi) \to \varphi$	(7,8 MP)
φ_{10}	$= \varphi$	(6,9 MP) □

Exercise 20

Consider the following axiom scheme, capturing "proof by contradiction"; call it (L3′):

$$(\neg\varphi \to \neg\psi) \to ((\neg\varphi \to \psi) \to \varphi).$$

Write $\Vdash \varphi$ to indicate that there is a deduction of φ using the axioms (L1), (L2), and (L3′) (instead of (L1)–(L3)).

(a) Show that $\vdash (\neg\varphi \to \neg\psi) \to ((\neg\varphi \to \psi) \to \varphi)$.

(b) Show that $\Vdash (\neg\psi \to \neg\varphi) \to (\varphi \to \psi)$.

(c) Prove that for all formulas φ, $\vdash \varphi$ iff $\Vdash \varphi$.

Solution: page 202

1.5 Soundness and completeness

We are ready to conclude the chapter on propositional logic with the core results of *soundness* and *completeness*, which link truth with provability. Thinking about the notion of deduction we defined previously, we can identify two desiderata,

- Everything we can deduce is a tautology.[76]
- We can deduce *all* the tautologies.[77]

The first is soundness, the second completeness.

Theorem 10 (Soundness) *For every formula φ, if $\vdash \varphi$ then $\vDash \varphi$.*

Proof As in the proof of the Deduction Theorem (Theorem 8), we use induction on the length of the deduction of φ. Specifically, we prove the following: for all $n \geq 1$, if $\varphi_1, \ldots, \varphi_n$ is a deduction, then $\vDash \varphi_n$.

For the base case $n = 1$ the deduction is only one line long, thus the formula being deduced, φ_1, is an instance of (L1), (L2), or (L3). So what we must prove here is that each of these are tautologies. This is straightforward to do using the truth table method introduced above (in fact, you already did the work in Exercise 10!).

Now suppose inductively that the result holds for any deduction of length $k \leq n$; we will show it also holds for deductions of length $n+1$. Let $\varphi_1, \ldots, \varphi_{n+1}$ be a deduction. If φ_{n+1} is an instance

[76] This can be thought of as a basic "safety" condition: deductions will *never* lead us to false statements.

[77] This is more aspirational: deduction is powerful enough to carry us to *any* truth.

of an axiom scheme, then as we checked above we are done. So suppose not; then φ_{n+1} must follow from two previous lines by modus ponens: that is, there are $i, j < n + 1$ with $\varphi_i = \varphi_j \to \varphi_{n+1}$.

By the inductive hypothesis, since both φ_i and φ_j have deductions that are at most n lines long,[78] we can conclude that $\vDash \varphi_j$ and $\vDash \varphi_j \to \varphi_{n+1}$. But this implies that $\vDash \varphi_{n+1}$ as well—if not, then for some valuation v we would have $[\![\varphi_{n+1}]\!]_v = false$, but since $\vDash \varphi_j$ we know $[\![\varphi_j]\!]_v = true$, from which it follows (using the recursive definition of $[\![\cdot]\!]_v$) that $[\![\varphi_j \to \varphi_{n+1}]\!]_v = false$, contradicting the fact that $\vDash \varphi_j \to \varphi_{n+1}$. $\square$

[78] By Lemma 7.

Aside from it being nice to know that deductions will never lead us astray from tautologies, the Soundness Theorem also provides a way to tell when a formula is *not* deducible—something which we had no clear way of doing up until now.

Consider, for example, the formula $p \to (p \to q)$; is this deducible? In fact it is not, but how can this be proved? We've seen that some deductions can be fairly long and convoluted.[79] So how can we be *sure* that there's not some very clever, hard-to-find deduction of $p \to (p \to q)$ from the axioms? We can use soundness!

[79] For example, Proposition 9!

> ## Exercise 21
>
> Show that $p \to (p \to q)$ is not deducible.
>
> Solution: page 203

Completeness is harder. We need to show that every tautology is deducible, or, equivalently, that anything that is not deducible is not a tautology. To do this, we will show how to construct, given a formula φ that is not deducible, a valuation under which φ is false. The machinery required for this construction begins with the concept of *consistency*.[80]

A set of formulas Γ is called **consistent** if, for every formula ψ, it is not the case that $\Gamma \vdash \psi$ and $\Gamma \vdash \neg\psi$. So intuitively, a set is consistent whenever you can't deduce a contradiction from it.

[80] Roughly speaking, we will argue as follows: (1) if $\nvdash \varphi$, then $\{\neg\varphi\}$ is consistent; (2) given any consistent set, there is a valuation that makes all the formulas in that set true; (3) therefore, if $\nvdash \varphi$, there is a valuation that makes $\neg\varphi$ true, hence $\nvDash \varphi$.

What kind of sets are consistent? We will see later a useful characterization, but at this stage it is worthwhile at least to observe that $\varnothing$ is consistent—that is, we can't deduce a contradiction just from the axioms.

Exercise 22

Prove that $\varnothing$ is consistent.

Hint: page 204

Moreover, if we start with any consistent set Γ and find a formula φ that cannot be deduced from Γ, adding the negation of φ to Γ preserves consistency.[81]

Proposition 11 *If $\Gamma \nvdash \varphi$, then $\Gamma \cup \{\neg\varphi\}$ is consistent.*

Proof Suppose not; then there exists a formula ψ such that $\Gamma \cup \{\neg\varphi\} \vdash \psi$ and $\Gamma \cup \{\neg\varphi\} \vdash \neg\psi$. By the Deduction Theorem (Theorem 8), this means $\Gamma \vdash \neg\varphi \rightarrow \psi$ and $\Gamma \vdash \neg\varphi \rightarrow \neg\psi$. Putting this together with Exercise 20(a), we obtain $\Gamma \vdash \varphi$, a contradiction. $\square$

[81] Notice that in the statement of this proposition, we don't actually insist that Γ be consistent. Why not? (Hint: consider Exercise 24.)

Exercise 23

Show in detail (i.e., by chaining together deductions in the appropriate way) why $\Gamma \vdash \varphi$ follows from Exercise 20(a) in the preceding proof.

Exercise 24

Prove the "principle of explosion": if Γ is not consistent, then for *every* formula φ, $\Gamma \vdash \varphi$. (You can either try to prove this directly, or use Proposition 11.)

Solution: page 204

Γ is called **maximally consistent** if Γ is consistent and for every proper extension $\Gamma' \supsetneq \Gamma$, Γ' is *not* consistent. In other words, a maximally consistent set is a consistent set that cannot be made any larger without rendering it inconsistent.

Maximally consistent sets have an alternative characterization that is useful for understanding and reasoning about them.

Proposition 12 *A set of formulas Γ is maximally consistent if and only if it is consistent and, for each formula φ, either $\varphi \in \Gamma$ or $\neg\varphi \in \Gamma$.*

Proof First suppose that Γ is consistent and, for each formula φ, Γ contains either φ or $\neg\varphi$. We must show that Γ is maximally consistent. Let $\Gamma' \supsetneq \Gamma$ be a proper extension of Γ. Then there is some formula $\psi \in \Gamma'$ that is not in Γ. By assumption, since $\psi \notin \Gamma$, we know that $\neg\psi \in \Gamma$. But this means that $\Gamma' \vdash \psi$ and $\Gamma' \vdash \neg\psi$,[82] so Γ' is inconsistent, as desired.

Conversely, suppose that Γ is maximally consistent; then of course it is consistent, so what we must show is that for any given φ, Γ contains either φ or $\neg\varphi$.

Suppose for contradiction that Γ contains neither φ nor $\neg\varphi$. Then $\Gamma \cup \{\varphi\}$ and $\Gamma \cup \{\neg\varphi\}$ are proper extensions of Γ and so by definition they must be inconsistent. This means that there is some formula ψ such that $\Gamma \cup \{\varphi\} \vdash \psi$ and also $\Gamma \cup \{\varphi\} \vdash \neg\psi$, and similarly, there is some formula χ such that $\Gamma \cup \{\neg\varphi\} \vdash \chi$ and $\Gamma \cup \{\neg\varphi\} \vdash \neg\chi$.

Applying the Deduction Theorem (Theorem 8) to the latter two, we see that $\Gamma \vdash \neg\varphi \rightarrow \chi$ and $\Gamma \vdash \neg\varphi \rightarrow \neg\chi$; putting this together with Exercise 20(a), we arrive at $\Gamma \vdash \varphi$. Applying the Deduction Theorem to the former two, we get $\Gamma \vdash \varphi \rightarrow \psi$ and $\Gamma \vdash \varphi \rightarrow \neg\psi$, which (since $\Gamma \vdash \varphi$) implies that $\Gamma \vdash \psi$ and $\Gamma \vdash \neg\psi$.[83] But this means that Γ is inconsistent, a contradiction! So Γ must indeed contain either φ or $\neg\varphi$. $\square$

[82] Each with a one-line deduction!

[83] As usual, by chaining together deductions.

Exercise 25

Show that maximally consistent sets are *deductively closed*: that is, if Γ is maximally consistent and $\Gamma \vdash \varphi$, then $\varphi \in \Gamma$.

Solution: page 204

Maximally consistent sets are *big*; Proposition 12 tells us (intuitively) that they have a "yes/no opinion" on every single formula.[84] Do such sets even exist? The answer is yes, as follows from the following crucial result.

Lemma 13 (Lindenbaum's Lemma) *Every consistent set Γ can be extended to a maximally consistent set $\Gamma' \supseteq \Gamma$.*

Proof The proof is a recursive construction passing through each formula one by one. So we begin with an enumeration of all the formulas: $\varphi_1, \varphi_2, \varphi_3, \ldots$.[85]

The idea of the recursive construction is to iteratively build Γ' by adding each φ_n to the set only if doing so does not create an inconsistency.

Formally, define $\Gamma_0 = \Gamma$; this is our starting point and the base case of the recursive construction. Next, recursively define

$$\Gamma_{n+1} = \begin{cases} \Gamma_n \cup \{\varphi_{n+1}\} & \text{if } \Gamma_n \cup \{\varphi_{n+1}\} \text{ is consistent} \\ \Gamma_n & \text{otherwise.} \end{cases}$$

So at each stage, we are either adding one formula to our set or adding nothing.[86] Define

$$\Gamma' = \bigcup_{n=0}^{\infty} \Gamma_n.$$

We will show that Γ' is a maximally consistent extension of Γ. There are three things to establish.

First, $\Gamma' \supseteq \Gamma$. This is obvious, since

$$\Gamma = \Gamma_0 \subseteq \Gamma_1 \subseteq \cdots \subseteq \Gamma_n \subseteq \cdots \subseteq \Gamma'.$$

Second, Γ' is consistent.[87] Suppose for contradiction that Γ' is not consistent, so there is a ψ such that $\Gamma' \vdash \psi$ and $\Gamma' \vdash \neg\psi$. Let $\psi_1, \ldots, \psi_m$ be a deduction of ψ from Γ', and let $\psi'_1, \ldots, \psi'_{m'}$ be a deduction of $\neg\psi$ from Γ'. Let $A = \Gamma' \cap \{\psi_1, \ldots, \psi_m, \psi'_1, \ldots, \psi'_{m'}\}$; then A is clearly finite, and moreover it includes all the formulas in Γ' that were actually *used* as assumptions to derive a contradiction. This is the key: only finitely many assumptions

[84] Intuitively, when $\varphi \in \Gamma$ that means Γ has a "yes" opinion on φ, and when $\neg\varphi \in \Gamma$ that means Γ has a "no" opinion on φ. In this terminology, Proposition 12 says exactly that a maximally consistent is "opinionated" about every formula.

[85] There are many ways to produce such an enumeration. For example, first write down all the formulas of length ≤ 1 involving only the primitive propositions in the set $\{p_1\}$. Then at stage k, write down all the formulas of length $\leq k$ involving only primitive propositions in the set $\{p_1, \ldots, p_k\}$. Each of these sets is finite. Note that it is essential here that the set of primitive propositions, PROP, is itself countable. See Appendix A.4.

[86] More to the point: at stage k we are adding formula φ_k if and only if doing so preserves consistency.

[87] This is the crux of the proof; it relies on the fact that our formal notion of deduction is fundamentally *finitistic* in the sense that all deductions are finite.

[88] Indeed, every deduction is
limited to using only finitely
many assumptions (since it
can only be finitely many
lines long). In this sense no
deduction ever truly relies on
an infinite set of
assumptions.

[89] If you're not convinced of
this, first consider the case
where A has only one
element. Then try two. Then
use induction!

[90] If it were, then $\psi = \varphi_k$ would
have been added to Γ_{k-1} at
stage k.

[91] Consider, for instance, the
extension of $\varnothing$ produced by
an enumeration that begins
with p versus one that
begins with $\neg p$.

were actually needed.[88] Although Γ' is an infinite set, it was constructed step by step, one formula at a time, so there must have been a *finite* stage N where all the assumptions from A had already been added, so that $\Gamma_N \supseteq A$.[89] But this then implies that $\psi_1, \ldots, \psi_m$ is a deduction of ψ from Γ_N, and $\psi'_1, \ldots, \psi'_{m'}$ is a deduction of $\neg\psi$ from Γ_N. Hence Γ_N is inconsistent, which is a contradiction since our construction began with a consistent set $\Gamma_0 = \Gamma$ and preserved consistency at every stage.

Third, and finally, we must show that Γ' is *maximally* consistent. We will show it has no consistent, proper extension. Suppose $\Gamma'' \supsetneq \Gamma'$; we wish to show that Γ'' is not consistent. Let $\psi \in \Gamma'' \setminus \Gamma'$. Then $\psi = \varphi_k$, for some k. Since $\psi \notin \Gamma'$, we know that $\varphi_k \notin \Gamma_k$, which by definition means that $\Gamma_{k-1} \cup \{\psi\}$ is not consistent.[90] This means we can deduce a contradiction from $\Gamma_{k-1} \cup \{\psi\}$, and so clearly we can also deduce a contradiction from Γ'', since $\Gamma'' \supseteq \Gamma_{k-1} \cup \{\psi\}$. Thus Γ'' is inconsistent, which completes the proof.   □

It's worth noting that the particular maximally consistent extension Γ' constructed in this proof depends on the enumeration of formulas we begin with—a different order for the enumeration can produce a different extension.[91]

Maximally consistent sets are important in the proof of completeness because they are rich enough that we can construct valuations out of them, and particularly well-behaved valuations at that. Maximally consistent sets are fundamentally *syntactic* objects: sets of formulas defined in terms of what can and cannot be deduced from them. Valuations, of course, are fundamentally *semantic*: they are the truth-givers. So the following lemma may be viewed as the key point of contact between these two worlds—syntactic and semantic, proof and truth—that will allows us to prove completeness.

Lemma 14 (Truth Lemma) *Let Γ be a maximally consistent set, and define a valuation $v_\Gamma : \text{PROP} \to \{true, false\}$ by*

$$v_\Gamma(p) = \begin{cases} true & \text{if } p \in \Gamma \\ false & \text{if } p \notin \Gamma. \end{cases}$$

Then for all formulas φ, we have

$$[\![\varphi]\!]_{v_\Gamma} = true \text{ iff } \varphi \in \Gamma.$$

Proof Induction on the structure of φ. For the base case we must show that

$$[\![p]\!]_{v_\Gamma} = true \text{ iff } p \in \Gamma;$$

since $[\![p]\!]_{v_\Gamma} = v_\Gamma(p)$, this is just the definition of v_Γ.

So suppose inductively the result holds for φ; we wish to show it holds for $\neg\varphi$. That is, we want to show that

$$[\![\neg\varphi]\!]_{v_\Gamma} = true \text{ iff } \neg\varphi \in \Gamma.$$

By the recursive definition of $[\![\cdot]\!]_{v_\Gamma}$, we know that

$$[\![\neg\varphi]\!]_{v_\Gamma} = true \text{ iff } [\![\varphi]\!]_{v_\Gamma} = false;$$

moreover, by the inductive hypothesis, we have

$$[\![\varphi]\!]_{v_\Gamma} = false \text{ iff } \varphi \notin \Gamma.$$

Thus to complete this inductive step it suffices to show that

$$\varphi \notin \Gamma \text{ iff } \neg\varphi \in \Gamma.$$

This is where we use the fact that Γ is maximally consistent. First suppose that $\neg\varphi \in \Gamma$; then clearly $\varphi \notin \Gamma$, since otherwise Γ would be inconsistent. Conversely, suppose that $\varphi \notin \Gamma$; then by maximality (and specifically, applying Proposition 12), we must have $\neg\varphi \in \Gamma$, as desired.

Now suppose inductively that the result holds for φ and ψ; we wish to show it holds for $\varphi \to \psi$. So we must show that

$$[\![\varphi \to \psi]\!]_{v_\Gamma} = true \text{ iff } \varphi \to \psi \in \Gamma.$$

As above, using the definition of $[\![\cdot]\!]_{v_\Gamma}$ and the inductive hypothesis, we obtain

$$[\![\varphi \to \psi]\!]_{v_\Gamma} = true \text{ iff } [\![\varphi]\!]_{v_\Gamma} = false \text{ or } [\![\psi]\!]_{v_\Gamma} = true$$
$$\text{iff } \varphi \notin \Gamma \text{ or } \psi \in \Gamma,$$

so we will be done if we can prove that

$$\varphi \notin \Gamma \text{ or } \psi \in \Gamma \text{ iff } \varphi \to \psi \in \Gamma.$$

Again, we use maximal consistency. First suppose that $\varphi \to \psi \in \Gamma$. If $\varphi \notin \Gamma$, we are done; otherwise, if $\varphi \in \Gamma$, then $\Gamma \vdash \psi$,[92] in

[92] Since $\{\varphi, \varphi \to \psi\} \vdash \psi$.

which case by Exercise 25 we have $\psi \in \Gamma$, so again we are done. Conversely, suppose that $\varphi \notin \Gamma$ or $\psi \in \Gamma$; obviously there are two cases to consider. First, if $\psi \in \Gamma$, then since $\vdash \psi \to (\varphi \to \psi)$ it's easy to see that $\Gamma \vdash \varphi \to \psi$, so by Exercise 25 we obtain $\varphi \to \psi \in \Gamma$. Second, if $\varphi \notin \Gamma$, then by maximality we have $\neg \varphi \in \Gamma$, so by Exercise 18(d) we must have $\Gamma \vdash \varphi \to \psi$; thus Exercise 25 once again yields $\varphi \to \psi \in \Gamma$. □

Exercise 26

What can go wrong if we try to define v_Γ as above using an *inconsistent* set Γ, or a consistent set that is not *maximally* consistent?

The Truth Lemma shows us, first, that a maximally consistent set Γ can be used to construct a valuation, v_Γ, and second, that this valuation has exactly the same "opinions" on formulas (interpreted via the recursive definition of truth) as Γ does. This at last puts us into position to prove completeness.

Theorem 15 (Completeness) *For every formula φ, if $\models \varphi$ then $\vdash \varphi$.*

Proof We prove the contrapositive. Suppose that $\nvdash \varphi$. Then by Proposition 11, $\{\neg\varphi\}$ is consistent. By Lindenbaum's Lemma, there exists a maximally consistent extension of $\{\neg\varphi\}$, call it Γ. Then by the Truth Lemma we have $[\![\neg\varphi]\!]_{v_\Gamma} = true$, from which it follows that $\nvDash \varphi$, as desired. □

In addition to the abstract benefits of having proved soundness and completeness,[93] this result also has an immediate, practical use: we never have to do a deduction again![94] This is because if at any point we need to show that some formula φ is deducible, Theorem 15 tells us that it's enough to simply draw up its truth table and check that it's a tautology.[95]

These results can be generalized to arbitrary sets of assumptions. Say that a valuation v **satisfies** a set of formulas Γ if, for all $\psi \in \Gamma$, $[\![\psi]\!]_v = true$, and call Γ **satisfiable** if there exists a

[93] Namely, knowing that deductions are both safe (never lead to untruths) and powerful (are capable of reaching any truth).

[94] At least not for propositional logic.

[95] If we weren't going to arrive at this result, it would have been more important to develop a more intuitive and human-friendly notion of deduction.

valuation that satisfies it. Say that Γ **entails** φ, and write $\Gamma \vDash \varphi$, if for every v that satisfies Γ we also have $[\![\varphi]\!]_v = true$.

Theorem 16 (Strong Soundness and Completeness) *For all sets of formulas Γ and all formulas φ, $\Gamma \vdash \varphi$ if and only if $\Gamma \vDash \varphi$.*

Exercise 27

Prove Theorem 16.

Solution: page 204

Exercise 28

Show that Γ is consistent if and only if it is satisfiable.

Solution: page 205

Exercise 29

Prove *compactness* of propositional logic: Γ is satisfiable if and only if every finite subset of Γ is satisfiable.

Solution: page 205

Exercise 30

Is it possible for a set to be consistent and closed under deduction, but not maximally consistent?

Exercise 31

Given a valuation v, define

$$P_v = \{p : v(p) = true\} \cup \{\neg q : v(q) = false\}.$$

(a) Show that P_v is consistent.
(b) Show that P_v has *exactly one* maximally consistent extension.
(c) Let Γ_v denote the unique maximally consistent extension of P_v. Show that for all maximally consistent sets Γ and all valuations v, we have $\Gamma_{v_\Gamma} = \Gamma$ and $v_{\Gamma_v} = v$.

(d) What can you conclude from part (c) about the relationship between the set of all valuations and the set of all
maximally consistent sets?

(e) Show that there are continuum-many maximally consistent sets.

Solution: page 206

2 Classical Predicate Logic

We now expand our attention to *predicate logic*,[1] which enriches propositional logic by adding *universal*[2] and *existential*[3] quantification. Quantification is absolutely ubiquitous in mathematics, from the statement of Goldbach's conjecture[4] to the definition of continuity,[5] to the very concept of soundness and completeness.[6] Beyond mathematics proper, it is also fundamental to the expression of countless concepts in everyday life, from settling on the best route home during rush hour[7] to planning a seating arrangement at a wedding.[8]

Expanding our attention to include quantification involves a lot more than just adding new symbols $\forall$ and $\exists$ to the language to stand for universal and existential quantification—we also need to introduce the things we are *quantifying over*, and give ourselves the tools to reason about them. To achieve this, we will radically alter the foundations of propositional logic to incorporate "objects" in addition to propositions, as well as ways of talking about and (sometimes) transforming these objects. We begin, as always, with language.

2.1 Language

To get a better sense of the building blocks we'll need for the language of predicate logic, let's consider a more formal expression of Goldbach's conjecture:

$$\forall x((E(x) \wedge x > 2) \rightarrow \exists y \exists z (P(y) \wedge P(z) \wedge x = y + z)).$$

This statement is intended to be read as follows: "For all numbers x, if x is even and x is greater than 2, then there exist numbers y and z such that y is prime and z is prime and the sum of y and z is x". There is a *lot* going on in this

---

[1] Also called *first-order logic*.

[2] "For every ..."

[3] "There exists ..."

[4] Every even number greater than 2 is the sum of two prime numbers.

[5] For every a and every $\varepsilon > 0$ there exists a $\delta > 0$ such that, for all x, if $|x - a| < \delta$ then $|f(x) - f(a)| < \varepsilon$.

[6] For every formula φ, φ is deducible if and only if φ is always true.

[7] There does not exist a faster route.

[8] For example, for every table there must exist two people at the table who already know each other.

single statement; analyzing its component parts and under-lying assumptions will give us some insight into what we need to carefully formalize.

- As noted previously, we are using the symbol ∀ to stand for universal quantification.[9]
- We need to quantify "over" something—in this particular case it is "numbers", though in general we may wish to reason about other domains.[10]
- We also need to be able to refer back to what we're quantifying over, somehow; we help ourselves to the *variable x* for this purpose.
- We are using $E(x)$ to stand for "x is even"; E is called a *predicate*.[11]
- The propositional connective ∧ is playing the same role it did in propositional logic, that is, it stands for "and". Predicate logic will essentially subsume and extend propositional logic.
- We are using the symbol ">" to stand for the relation of "greater than" between numbers.[12]
- We allow ourselves access to the symbol "2" as a way of denoting a particular number; this is called a *constant* (in contrast with a variable).
- The statement of Goldbach's conjecture has the form "Every number with property A has property B",[13] which is expressed using material implication (denoted, as usual, with the symbol →): for all x, *if* x has property A, *then* it has property B.
- The statement "x is the sum of two primes" contains some hidden quantification, this time of the existential variety: *there exist* two numbers that are prime and sum to x. We use the symbol ∃ to stand for existential quantification,[14] and as previously help ourselves to two new variables to keep track what we're existentially quantifying.
- Like with the predicate E for "is even", we use $P(x)$ to stand for "x is prime".[15]
- We are allowing ourselves access to the operation of addition; unlike a predicate or relation, this does not produce a statement but rather a new number.[16]
- Finally, we are using the "=" symbol to express equality.[17]

[9] It is an upside-down "A", which stands for "all".

[10] For example, sets, functions, traffic routes, X (formerly Twitter) accounts—anything we can quantify, really.

[11] One could also imagine replacing this predicate with an expression like $\exists w(x = 2w)$, which seems like a more explicit way of capturing x being even. More on this later.

[12] This can also be viewed as a sort of "two-place" predicate; see sidenote 24.

[13] Here, of course, property A is being even and greater than 2, while property B is being the sum of two primes.

[14] It is a back-to-front "E", which stands for "exists".

[15] Think about whether this predicate can also be expressed more explicitly, using quantification.

[16] For example: 2 < 3 is a statement that we can judge to be true or false; by contrast, 2 + 3 is neither true nor false, but rather a way of expressing a new number.

[17] Like >, this can be viewed as a binary relation.

This (long!) list contains a mixture of both general-purpose tools we'll want available for all first-order languages[18] together with more specific tools that are useful in the particular context of talking about something like Goldbach's conjecture.[19] To specify the language of predicate logic in a fully general way, we will not insist that it contain any particular fixed list of predicates, functions, and constants, but rather, allow it to flexibly include whichever such symbols we wish to have access to, in a way that can vary from application to application.

This leads us formalize the notion of a *signature*:[20]

$$S = (f_1, \ldots, f_n, r_1, \ldots, r_m, \{\bar{c}_i\}_{i \in I}).$$

A signature consists of finitely many *function symbols* $f_1, \ldots, f_n$, finitely many *relation symbols* $r_1, \ldots, r_m$,[21] and a set (not necessarily finite) of *constant symbols* $\bar{c}_i$.[22] Each function and relation symbol is assumed to come equipped with a positive natural number called its *arity*, which intuitively is the number of arguments it is intended to accept. To avoid clutter we do not indicate arities explicitly in the notation.[23]

Intuitively, these symbols are meant to be interpreted over some *domain of discourse* consisting of the objects we wish to reason about. So function symbols are meant to be interpreted as actual functions on this domain, predicate symbols as actual predicates (i.e., relations) on this domain,[24] and constants as actual elements of this domain.

As a concrete example, we might wish to reason about the natural numbers—about things like adding and multiplying, primeness, and comparative size. We can do this using a signature such as $S = (a, m, p, \ell, \{\bar{0}, \bar{1}\})$ interpreted over the domain of natural numbers, $\mathbb{N}$, with:

[18] For example, quantifiers, variables, and logical connectives.

[19] For example, predicates for being even and prime, the greater-than relation, the number 2, and the operation of addition.

[20] The set I here is just an "index set", meant to keep track of all the possible constant symbols (there might be a lot, even uncountably many!). Don't pay attention to it for now if it's confusing.

[21] Also called *predicate symbols*.

[22] The reason for writing the constant symbols with a bar over top will become clear shortly.

[23] For instance, some textbooks note the arity of each function symbol and predicate symbol as a superscript, so, for example, if f_i had arity k it would instead be written f_i^k.

[24] As we have seen, a "predicate" is a description that may apply to objects, pairs of objects, triples, etc. For example, earlier we considered the unary predicate "is prime", which applies to individual numbers, and the binary predicate "is greater than", which applies to pairs of numbers. The former essentially picks out a subset of $\mathbb{N}$: the set of prime numbers. The latter picks out a subset of $\mathbb{N} \times \mathbb{N}$: all pairs (n, m) such that n is greater than m. This is why we can also think of predicates as relations. See Appendix A.1.

[25] So a is a function symbol of arity 2.

[26] Similarly m is a function symbol of arity 2.

[27] This means p is a predicate symbol of arity 1.

[28] And ℓ is a predicate symbol of arity 2.

[29] Notice here how it's useful to have the constant symbols $\bar{0}$ and $\bar{1}$ notationally "marked" with the bar, so we can keep them distinct from the actual numbers 0 and 1.

[30] Notice also that there are many, many more objects in our domain of discourse than there are constant symbols to "name" them. There is no requirement that we have a constant symbol for every object.

- $a(x,y)$ interpreted as "$x + y$",[25]
- $m(x,y)$ as "$x \cdot y$",[26]
- $p(x)$ as "x is prime",[27]
- $\ell(x,y)$ as "$x < y$",[28]
- $\bar{0}$ as "0", and
- $\bar{1}$ as "1".[29,30]

We will make this idea of "interpretation" precise in Section 2.2 on semantics. For now it is useful simply to hold it in mind for the sake of intuition—to give some concrete idea of how we will ultimately try to *use* the formal languages we construct.

It's important to remember that just as in propositional logic, before any notion of semantics or deduction has been provided, the symbols themselves are officially meaningless. We may have *in mind* an intended interpretation, but it has not yet been defined or formalized. Aside from the importance of formally distinguishing a logical language from its interpretation, this approach also provides the flexibility we sought previously. Recall that in propositional logic we began with an abstract collection PROP = $\{p_1, p_2, \ldots\}$ of primitive propositions precisely because we wanted our formalism to be versatile—we didn't want to pre-commit ourselves to a language appropriate for reasoning about just one topic, but rather be able to interpret it in all sorts of ways. The same thing is happening here: in the example we have just seen, although we *can* interpret the symbol a as addition, m as multiplication, etc., over the natural numbers, nothing prevents us from interpreting the very same symbols in totally different ways, perhaps over a totally different domain.

[31] Note that this means there are many languages of predicate logic—one for each signature!

A signature *generates a language* in two stages.[31] First we define the **terms** of the language recursively by "applying" function symbols to constants and to *variables*, the latter of which we always assume are drawn from some fixed, countably infinite set VAR = $\{x_1, x_2, \ldots\}$.

- Each constant $\bar{c}$ and each variable x is a term.
- If $t_1, \ldots, t_n$ are terms and f is a function symbol of arity n, then $f(t_1, \ldots, t_n)$ is a term.

Formulas are then built recursively from terms as follows.

- If $t_1, \ldots, t_n$ are terms and r is a relation symbol of arity n, then $r(t_1, \ldots, t_n)$ is a formula.[32]
- If t_1 and t_2 are terms, then $t_1 = t_2$ is a formula.[33]
- If φ and ψ are formulas, then so are $\neg\varphi$ and $\varphi \to \psi$.[34]
- If φ is a formula and x is a variable, then $\forall x\varphi$ and $\exists x\varphi$ are formulas too.[35,36]

Roughly speaking, terms are the syntactic structures that are intended to be interpreted as *elements* of the domain of discourse, while formulas are intended to bear truth values. Formulas of the form $r(t_1, \ldots, t_n)$ or $t_1 = t_2$ are analogues of primitive propositions: each such formula will ultimately be either true or false (in a given interpretation), and all other formulas are constructed from these basic building blocks using connectives and quantifiers. So we have at our disposal all the standard connectives of propositional logic plus universal and existential quantification.

Consider, for example, the signature with relation symbols f and b of arity 2 and 1, respectively, and a constant symbol $\bar{t}$. Assume the domain of discourse is the set of all X (formerly Twitter) accounts, the relation $f(x, y)$ means "x follows y", the predicate $b(x)$ means "x is a bot", and the constant symbol $\bar{t}$ denotes the official X account (@X). Then we can use the first-order language generated by this signature to formalize a variety of statements about X. For instance, we can state the following.

- No account follows itself:[37]

$$\forall x \neg f(x, x) \ (\text{or: } \neg \exists x f(x, x)).$$

- There is a bot that follows the official X account:[38]

$$\exists x (b(x) \wedge f(x, \bar{t})).$$

- Every bot follows the official X account:[39]

$$\forall x (b(x) \to f(x, \bar{t})).$$

[32] Henceforth we assume that whenever we write an expression like $f(t_1, \ldots, t_n)$ or $r(t_1, \ldots, t_n)$, the arity of the function/relation symbol matches the number of arguments.

[33] We could have included this in the previous clause simply by including an "equality relation" (of arity 2) in the signature. We include it here instead to emphasize that we will always (unless otherwise noted) assume our language has equality in it, no matter what the signature is.

[34] Other binary connectives can be treated as abbreviations as in Chapter 1.

[35] These are read, "for all x, φ" and "there exists an x such that φ", respectively.

[36] As in the propositional case, we insert parentheses where needed for disambiguation. By convention, quantifiers are assumed to take narrow scope (like negation).

[37] For all X accounts x, it is not the case that x follows x (or: it is not the case that there is an X account x such that x follows x).

[38] There exists an X account x such that x is a bot and x follows @X.

[39] For all X accounts x, if x is a bot, then x follows @X.

[40] There is an X account x such that @X follows x and moreover, for any X account y, if @X follows y then y is x.

[41] For all X accounts x and y, if x follows y and y follows x, then there is an X account z such that x follows z and y follows z.

[42] For all X accounts x, if there is an X account y such that y is a bot and x follows y, then there is an X account z such that x follows z and z is not y.

[43] It is not the case that there is an X account x such that, for all X accounts y and z, if x follows y and y follows z, then x follows z.

- The official X account follows exactly one account:[40]

$$\exists x(f(\bar{t},x) \wedge \forall y(f(\bar{t},y) \to y = x)).$$

- Whenever two accounts follow each other, there is an account they follow in common:[41]

$$\forall x\forall y((f(x,y) \wedge f(y,x)) \to \exists z(f(x,z) \wedge f(y,z))).$$

- Any account that follows a bot follows at least one other account:[42]

$$\forall x(\exists y(b(y) \wedge f(x,y)) \to \exists z(f(x,z) \wedge \neg z = y)).$$

- No account follows all the accounts that are followed by the accounts it follows:[43]

$$\neg\exists x\forall y\forall z((f(x,y) \wedge f(y,z)) \to f(x,z)).$$

Exercise 32

Consider a signature with a binary relation symbol ℓ and a constant symbol $\bar{0}$. Assume the domain of discourse is $\mathbb{N}$, the relation $\ell(x,y)$ means "$x < y$", and the constant symbol $\bar{0}$ denotes the number 0. Formalize the following statements in the language generated by this signature.

(a) 0 is the smallest number.
(b) There is a smallest number.
(c) There is no largest number.
(d) Every number has an immediate successor (i.e., for every number, there one that is "next biggest" in the ordering).
(e) Every number is greater than some other number.

Now add to the signature binary functions a and m (interpreted as addition and multiplication, respectively), and the constant symbol $\bar{1}$ (interpreted as the number 1), and formalize the following.

(g) All square numbers are positive.

(h) There's exactly one even prime number.

(i) Between every two squares there is always a prime.

Solution: page 207

Exercise 33

Let $S = (f)$, where f is a function symbol of arity 1. Describe the set of terms in the language generated by S.

Solution: page 207

As in the case of propositional logic, we can take advantage of the recursive structure of the language to make recursive definitions. To begin with a simple example, we define the set $FV(\varphi)$ of *free variables* of φ, that is, the set of all variables occurring in φ that are not "bound" by a quantifier.[44] Since terms and formulas are each defined recursively, our recursive definition must proceed through the base cases and recursive constructions of each:

[44] A variable x is called *bound* in φ if it occurs inside the scope of the universal quantifier $\forall x$ or the existential quantifier $\exists x$. See Exercise 34.

- $FV(\overline{c}) = \varnothing$
- $FV(x) = \{x\}$
- $FV(f(t_1,\ldots,t_n)) = FV(t_1) \cup \cdots \cup FV(t_n)$
- $FV(r(t_1,\ldots,t_n)) = FV(t_1) \cup \cdots \cup FV(t_n)$
- $FV(t_1 = t_2) = FV(t_1) \cup FV(t_2)$
- $FV(\neg\varphi) = FV(\varphi)$
- $FV(\varphi \rightarrow \psi) = FV(\varphi) \cup FV(\psi)$
- $FV(\forall x\varphi) = FV(\exists x\varphi) = FV(\varphi) \setminus \{x\}$.

This definition is meant to reflect the simple idea that any variable that occurs anywhere in a term or formula occurs free *unless* it falls within the scope of a quantifier that binds it.[45]

[45] It follows easily that in a term *all* variables are free, since terms cannot contain quantifiers.

Exercise 34

(a) Give a recursive definition of $BV(\varphi)$, the set of *bound variables* of φ.

(b) Is it always the case that $FV(\varphi) \cap BV(\varphi) = \varnothing$? Justify your answer with a proof or counterexample.

Solution: page 208

A term or formula is called **closed** if it has no free variables. A closed formula is also called a *sentence*. Since the whole point of adding variables to our language is to enable quantification, we will be especially interested in sentences, since in this case all the variables are "doing their job". Indeed, in Section 2.2 we will see that only closed formulas get assigned truth values.[46]

Another important concept we can define recursively is the *substitution of t for x in* φ, denoted $\varphi[t/x]$. One might think this is simply the result of replacing each occurrence of the variable x in the formula φ with the term t, but this is not quite right: we only want to substitute t for *free* occurences of x in φ. The recursive definition that follows reflects this:

- $\bar{c}[t/x] = \bar{c}$
- $y[t/x] = \begin{cases} t & \text{if } y \text{ is } x \\ y & \text{otherwise} \end{cases}$
- $f(t_1, \ldots, t_n)[t/x] = f(t_1[t/x], \ldots, t_n[t/x])$
- $r(t_1, \ldots, t_n)[t/x] = r(t_1[t/x], \ldots, t_n[t/x])$
- $(t_1 = t_2)[t/x] = (t_1[t/x]) = (t_2[t/x])$[47]
- $(\neg\varphi)[t/x] = \neg(\varphi[t/x])$
- $(\varphi \to \psi)[t/x] = (\varphi[t/x]) \to (\psi[t/x])$
- $(\forall y\varphi)[t/x] = \begin{cases} \forall y\varphi & \text{if } y \text{ is } x \\ \forall y(\varphi[t/x]) & \text{otherwise} \end{cases}$
- $(\exists y\varphi)[t/x] = \begin{cases} \exists y\varphi & \text{if } y \text{ is } x \\ \exists y(\varphi[t/x]) & \text{otherwise.} \end{cases}$

As promised, the last two clauses block substitution from progressing for any variable within the scope of a quantifier of

[46] To foreshadow a bit: it is difficult to interpret a formula like $x = y$ as being true or false, since intuitively x and y don't stand for specific objects (like constants do) but rather have a *variable* interpretation. By contrast, the closed formula $\forall x\forall y(x = y)$ does have a definite meaning—it says that there is only one object! So this would be false in any context where there is more than one object in the domain of discourse.

[47] Here we see why some authors prefer to use a different symbol for equality in the object language (e.g., $\doteq$, $\equiv$, etc.), in which case this recursive clause would be written instead as

$$(t_1 \doteq t_2)[t/x] = (t_1[t/x])$$
$$\doteq (t_2[t/x]).$$

Rather than introducing yet more notation, we instead rely on context to distinguish object language versus metalanguage uses of the "=" symbol.

that same variable. Thus, for example, if φ is the formula $\forall y(x = y)$, then $\varphi[\bar{c}/x] = \forall y(\bar{c} = y)$, but on the other hand, $\varphi[\bar{c}/y] = \forall y(x = y)$.[48]

For convenience, and following a widespread convention, we will often write $\varphi(x)$ to denote the formula φ with the variable x "marked", precisely so that we can write $\varphi(t)$ as a shorthand for $\varphi[t/x]$.[49]

Formulas with free variables can be thought of as "compound predicates" (with arity equal to the number of free variables). For instance, using the sample signature and interpretation from above, if the formula $\varphi(x)$ denotes

$$\ell(\bar{1}, x) \wedge \forall y \forall z(m(y, z) = x \rightarrow (y = \bar{1} \vee z = \bar{1})),$$

then $\varphi(x)$ expresses exactly the conditions for x being prime. It says, intuitively, "x is greater than 1 and whenever any two numbers multiplied together equal x, at least of them must be 1". The formula

$$\forall x(\varphi(x) \leftrightarrow p(x))$$

should therefore come out as true under this interpretation.[50] Of course, we have yet to define formally what it means for a formula of predicate logic to be "true". We turn to this task next.

2.2 Semantics: interpretation in structures

Intuitively, to give meaning to the terms and formulas of a first-order language, we first need to say what the basic objects are (over which we're quantifying), and also how to interpret all the specific symbols in the signature. Given a signature

$$\mathcal{S} = (f_1, \ldots, f_n, r_1, \ldots, r_m, \{\bar{c}_i\}_{i \in I}),$$

a **model** for $\mathcal{S}$ is a pair $\mathcal{A} = (A, [\![\cdot]\!]_\mathcal{A})$, where A is a non-empty set called the *domain of discourse*, and for each function symbol f, relation symbol r, and constant symbol $\bar{c}$ in $\mathcal{S}$ we have

$$[\![f]\!]_\mathcal{A} : A^k \rightarrow A \quad \text{(where k is the arity of f)},$$
$$[\![r]\!]_\mathcal{A} \subseteq A^\ell \quad \text{(where ℓ is the arity of r)},$$
$$[\![\bar{c}]\!]_\mathcal{A} \in A.$$

[48] If we did allow substitution for bound variables, it could "decouple" that variable from its quantifier. For instance, in the earlier example, we would have $\varphi[\bar{c}/y] = \forall y(x = \bar{c})$, in which case the $\forall y$ quantifier is no longer quantifying anything. While the result is still a well-formed formula, it will be much more convenient to work with a notion of substitution that ignores bound occurences of variables.

[49] A related (though slightly vexing) convention is that we do *not* assume that a formula written in this way with x marked necessarily contains any occurrence of the variable x. While this may seem perverse, we will see that it is sometimes useful to remain "agnostic" about whether a formula $\varphi(x)$ actually contains x or not.

[50] In particular, this means that even if we removed the predicate p from the language, we could still "capture" primeness using the other resources of the language.

[51] Notice that for each
signature S, there is an
immense amount of free
choice involved in specifying
a model: first one must
choose the domain of
discourse (which can be any
non-empty set whatsoever),
and then interpretations for
all the function, relation, and
constant symbols. This is the
flexibility we discussed in
Section 2.1: the language
does not force any particular
interpretation on us.

We call $[\![\cdot]\!]_A$ an *interpretation*: essentially, it is a function which takes as input symbols in the signature and produces as output the actual mathematical objects those symbols are meant to refer to.[51] These outputs are sometimes called *denotations*: thus, according to the interpretation $[\![\cdot]\!]_A$, the function symbol f denotes the *actual function* $[\![f]\!]_A : A^k \to A$, the relation symbol r denotes the *actual relation* $[\![r]\!]_A \subseteq A^\ell$, and the constant symbol $\bar{c}$ denotes the *actual element* $[\![\bar{c}]\!]_A \in A$.

This domain of discourse A taken together with all the functions, relations, and distinguished elements given by the denotations constitutes a mathematical structure sufficiently rich to interpret all the symbols in S. Formally, if A is a model for S, then the tuple

$$(A, [\![f_1]\!]_A, \ldots, [\![f_n]\!]_A, [\![r_1]\!]_A, \ldots, [\![r_m]\!]_A, \{[\![\bar{c}_i]\!]_A\}_{i \in I})$$

is called an S-**structure**. By a minor abuse of notation, we use the same symbol, A, for both a model and the corresponding S-structure.

Naturally, we want to extend this interpretation to all (closed) terms and formulas in the language. For example, if f is a function symbol of arity 1 and $\bar{c}$ is a constant symbol, then $f(\bar{c})$ is a (closed) term, and $[\![f(\bar{c})]\!]_A$ will denote the result of applying the actual function $[\![f]\!]_A : A \to A$ to the element $[\![\bar{c}]\!]_A$, namely, $[\![f]\!]_A([\![\bar{c}]\!]_A)$, which is itself an element of A. Intuitively, all closed terms should denote elements of A.

[52] So $[\![r]\!]_A$ is unary relation,
that is, a subset of
A—intuitively consisting of
those elements of A that
satisfy the predicate.

Similarly, if r is a relation symbol of arity 1,[52] and $\bar{c}$ is a constant symbol, then $r(\bar{c})$ is a (closed) formula, and $[\![r(\bar{c})]\!]_A$ will denote *true* if and only if $[\![\bar{c}]\!]_A \in [\![r]\!]_A$. This may feel somewhat less intuitive—what does it mean to "denote" a truth value? Perhaps a more natural phrasing is simply to say that all closed formulas ought to be *interpreted* as being either true or false. This is exactly the project of this section.

Our main goal here is to provide a recursive definition of truth for all closed formulas. To accomplish this, we will need to specify how to recursively define the truth of quantified formulas like $\forall x \varphi$ and $\exists x \varphi$. But how can we tell whether $\forall x \varphi$ is true *recursively*, that is, based on whether φ is true? Does this even make sense? After all, if φ contains x as a free variable, then it is not closed and intuitively has no truth value.

Of course, the idea is to interpret universal quantification as saying something about *all* values of x—for $\forall x \varphi$ to be true, we want to demand that φ is true no matter what x happens to be. How can we formalize this? One way is by radically expanding the collection of constant symbols in our signature, which will allow us to refer to every single object in the domain of discourse directly.

Given S and $\mathcal{A}$, we define the **extended signature** $S(\mathcal{A})$ to be the signature obtained from S by adding new constant symbols $\tilde{a}$, one for each $a \in A$. Intuitively, each new constant $\tilde{a}$ will denote the actual element $a \in A$.[53] This is made precise by the following definition, in which the interpretation $[\![\cdot]\!]_{\mathcal{A}}$ is extended recursively to all closed terms and closed formulas of $S(\mathcal{A})$:

- $[\![c_i]\!]_{\mathcal{A}} = c_i$
- $[\![\tilde{a}]\!]_{\mathcal{A}} = a$
- $[\![f(t_1,\ldots,t_n)]\!]_{\mathcal{A}} = [\![f]\!]_{\mathcal{A}}([\![t_1]\!]_{\mathcal{A}},\ldots,[\![t_n]\!]_{\mathcal{A}})$
- $[\![r(t_1,\ldots,t_n)]\!]_{\mathcal{A}} = \begin{cases} true & \text{if } ([\![t_1]\!]_{\mathcal{A}},\ldots,[\![t_n]\!]_{\mathcal{A}}) \in [\![r]\!]_{\mathcal{A}} \\ false & \text{otherwise} \end{cases}$
- $[\![t_1 = t_2]\!]_{\mathcal{A}} = \begin{cases} true & \text{if } [\![t_1]\!]_{\mathcal{A}} = [\![t_2]\!]_{\mathcal{A}}\,^{54} \\ false & \text{otherwise} \end{cases}$
- $[\![\neg\varphi]\!]_{\mathcal{A}} = \begin{cases} true & \text{if } [\![\varphi]\!]_{\mathcal{A}} \neq true \\ false & \text{otherwise} \end{cases}$
- $[\![\varphi \to \psi]\!]_{\mathcal{A}} = \begin{cases} true & \text{if } [\![\varphi]\!]_{\mathcal{A}} = false \text{ or } [\![\psi]\!]_{\mathcal{A}} = true \\ false & \text{otherwise} \end{cases}$
- $[\![\forall x \varphi]\!]_{\mathcal{A}} = \begin{cases} true & \text{if for all } a \in A,\ [\![\varphi[\tilde{a}/x]]\!]_{\mathcal{A}} = true \\ false & \text{otherwise} \end{cases}$
- $[\![\exists x \varphi]\!]_{\mathcal{A}} = \begin{cases} true & \text{if for some } a \in A,\ [\![\varphi[\tilde{a}/x]]\!]_{\mathcal{A}} = true \\ false & \text{otherwise.} \end{cases}$

[53] So some of these new constant symbols may end up denoting objects already denoted by constant symbols in S. In general, though, there is no guarantee that S has a separate constant symbol for each object in the domain of discourse. This is what the extended signature guarantees.

[54] Again, we see the two different uses of "=": the sentence $t_1 = t_2$ is true iff the objects denoted by the terms t_1 and t_2, namely $[\![t_1]\!]_{\mathcal{A}}$ and $[\![t_2]\!]_{\mathcal{A}}$, are the same.

Thus we see that constant, function, and relation symbols are interpreted as actual elements, functions, and relations (as outlined previously); the equality symbol is interpreted as actual equality; and the propositional connectives ($\neg$ and $\rightarrow$) are interpreted just as they were in propositional logic. Last but certainly not least, the clauses for universal and existential quantification are interpreted using the metalanguage concepts of "all" and "some" applied to the actual domain of discourse A, which is imported into the object language through all the new constant symbols added in the extended signature.

> ## Exercise 35
>
> Prove that for all closed terms t of $\mathcal{S}(A)$, $[\![t]\!]_A \in A$.

Let's examine the truth clauses for quantification a little more closely. By definition, the (closed) formula $\forall x \varphi$ is true (according to A) precisely when, for all $a \in A$, the formula $\varphi[\tilde{a}/x]$ is true (according to A). As promised, then, we are interpreting universal quantification by demanding that something be true for *all* elements a in the domain of discourse A. The formula $\varphi[\tilde{a}/x]$ is simply the result of starting with φ and replacing each free occurrence of x with $\tilde{a}$. Since $\tilde{a}$ is, by definition, interpreted as denoting the element $a \in A$, this provides the desired connection to the domain of discourse.[55] The clause for existential quantification is entirely analogous, swapping "all" with "some".

As usual when it comes to internalizing new definitions, practice makes perfect.

Proposition 17 *Let* $\mathcal{S} = (f, r, \{\bar{c}\})$, *where f is a function symbol of arity 2 and r is a relation symbol of arity 2, and consider the $\mathcal{S}$-structure* $A = (\mathbb{Z}, -, <, \{0\})$, *where* $[\![f]\!]_A = -$ *(i.e., subtraction),* $[\![r]\!]_A = <$, *and* $[\![\bar{c}]\!]_A = 0$. *Let φ be the formula*

$$\forall x \forall y (r(f(x,y), \bar{c}) \rightarrow r(x,y)).$$

Then $[\![\varphi]\!]_A = true$.

[55] Moreover, since $\forall x \varphi$ is assumed to be closed, it's easy to see that φ can have at most one free variable, namely x. Therefore $\varphi[\tilde{a}/x]$ must again be a closed formula, since each free occurrence of x has been replaced by the constant $\tilde{a}$. This ensures that it makes sense to consider the interpretation $[\![\varphi[\tilde{a}/x]]\!]_A$, which is only defined on sentences.

Proof We apply the recursive definition of truth to φ step by step, beginning of course with the outermost universal quantifiers, then the implication, and so on:[56]

$\llbracket \varphi \rrbracket_A = true$ iff for all $n \in \mathbb{Z}$, $\llbracket \forall y(r(f(\tilde{n},y),\bar{c}) \rightarrow r(\tilde{n},y)) \rrbracket_A = true$

iff for all $n,m \in \mathbb{Z}$, $\llbracket (r(f(\tilde{n},\tilde{m}),\bar{c}) \rightarrow r(\tilde{n},\tilde{m})) \rrbracket_A = true$

iff for all $n,m \in \mathbb{Z}$, $\llbracket (r(f(\tilde{n},\tilde{m}),\bar{c}) \rrbracket_A = true$

implies $\llbracket r(\tilde{n},\tilde{m}) \rrbracket_A = true$

iff for all $n,m \in \mathbb{Z}$, $(\llbracket f(\tilde{n},\tilde{m}) \rrbracket_A, \llbracket \bar{c} \rrbracket_A) \in \llbracket r \rrbracket_A$

implies $(\llbracket \tilde{n} \rrbracket_A, \llbracket \tilde{m} \rrbracket_A) \in \llbracket r \rrbracket_A$

iff for all $n,m \in \mathbb{Z}$, $\llbracket f \rrbracket_A(\llbracket \tilde{n} \rrbracket_A, \llbracket \tilde{m} \rrbracket_A) < 0$ implies $n < m$

iff for all $n,m \in \mathbb{Z}$, if $n - m < 0$ then $n < m$.

Since this is true, we have shown that $\llbracket \varphi \rrbracket_A = true$. $\qquad\square$

[56] Naturally, once you get the hang of it, it's much shorter to collapse several of these steps into one.

Exercise 36

Let S, A, and φ be defined as in Proposition 17. Define another S-structure B such that $\llbracket \varphi \rrbracket_A \neq \llbracket \varphi \rrbracket_B$.

Solution: page 208

Exercise 37

Let S and A be defined as in Proposition 17. Evaluate the following.

(a) $\llbracket \forall x(r(f(\bar{c},x),\bar{c})) \rrbracket_A$
(b) $\llbracket \forall x \forall y \forall z(r(x,y) \rightarrow r(f(x,z),f(y,z))) \rrbracket_A$
(c) $\llbracket \forall x \exists y(f(x,y) = \bar{c}) \rrbracket_A$
(d) $\llbracket \exists y \forall x(f(x,y) = \bar{c}) \rrbracket_A$

Solution: page 209

Exercise 38

Let $S = (s,a,m,\{\bar{0}\})$, where s, a, and m are function symbols of arities 1, 2, and 2, respectively, and consider the S-structure $\mathcal{N} = (\mathbb{N}, S, +, \cdot, \{0\})$, where $\llbracket s \rrbracket_{\mathcal{N}} = S$ (i.e., the successor function, defined by $S(x) = x + 1$), $[\![a]\!]_\mathcal{N} = +$, $[\![m]\!]_\mathcal{N} = \cdot$, and $[\![\bar{0}]\!]_\mathcal{N} = 0$.

(a) Give two distinct terms t in the language generated by S such that $[\![t]\!]_\mathcal{N} = 5$.

(b) Show that for each $n \in \mathbb{N}$, there is a term t in the language generated by S such that $[\![t]\!]_\mathcal{N} = n$.

(c) Show that for each $n \in \mathbb{N}$ there are infinitely many terms t in the language generated by S such that $[\![t]\!]_\mathcal{N} = n$.

Solution: page 209

Exercise 39

Let S and $\mathcal{N}$ be as defined in Exercise 38, and evaluate the following.

(a) $[\![\forall x(m(x, s(\bar{0})) = x)]\!]_\mathcal{N}$
(b) $[\![\forall x \forall y(a(x, \bar{0}) = y \rightarrow a(y, \bar{0}) = x)]\!]_\mathcal{N}$
(c) $[\![\forall x \forall y \exists z(a(x, z) = y)]\!]_\mathcal{N}$
(d) $[\![\exists x \forall y(m(x, y) = y))]\!]_\mathcal{N}$
(e) $[\![\forall x \forall y(m(x, s(y)) = a(m(x, y), x))]\!]_\mathcal{N}$
(f) $[\![\exists x(a(s(x), s(x)) = m(s(x), s(x)))]\!]_\mathcal{N}$
(g) $[\![\forall x(\neg(x = s(s(\bar{0}))) \rightarrow (\neg(m(x, x) = a(x, x))))]\!]_\mathcal{N}$

Solution: page 209

Exercise 40

Given any S-structure $\mathcal{A}$ and any closed term t, prove that

$$[\![t = \overline{[\![t]\!]_\mathcal{A}}]\!]_\mathcal{A} = true.$$

Solution: page 210

We sometimes write $\mathcal{A} \vDash \varphi$ to mean $[\![\varphi]\!]_\mathcal{A} = true$; in either case we say "φ is true in $\mathcal{A}$", or sometimes "$\mathcal{A}$ satisfies φ". A sentence φ in the language generated by S is called **valid** if it is true in *all* S-structures.[57] In this case we write $\vDash \varphi$.

[57] Equivalently, if it is true for all models.

Valid sentences are the analogues of tautologies from propositional logic: they are sentences that are true regardless of the specific, contingent interpretation of the symbols of the language. For example, if p is a relation symbol of arity 1 and $\bar{c}$ is a constant, then whether or not $p(\bar{c})$ is true clearly depends on the particular interpretation of p and $\bar{c}$.[58] However, one can easily check that a sentence like $\forall x p(x) \to p(\bar{c})$ comes out as true no matter what model we consider.[59]

Proposition 18 *Let S be a signature that includes a relation symbol p of arity 1 and a constant $\bar{c}$. Then $\vDash \forall x p(x) \to p(\bar{c})$. On the other hand, $\nvDash \forall x(p(x) \to p(\bar{c}))$.*

Proof First we must show that whenever $\mathcal{A} = (A, [\![\cdot]\!]_{\mathcal{A}})$ is a model for S, $\mathcal{A} \vDash \forall x p(x) \to p(\bar{c})$. Let's unravel what this means by using the recursive definition of truth:

$\mathcal{A} \vDash \forall x p(x) \to p(\bar{c})$ iff $\mathcal{A} \vDash \forall x p(x)$ implies[60] $\mathcal{A} \vDash p(\bar{c})$

$\qquad$ iff (for all $a \in A$, $\mathcal{A} \vDash p(\tilde{a})$) implies $[\![\bar{c}]\!]_{\mathcal{A}} \in [\![p]\!]_{\mathcal{A}}$

$\qquad$ iff (for all $a \in A$, $[\![\tilde{a}]\!]_{\mathcal{A}} \in [\![p]\!]_{\mathcal{A}}$) implies $[\![\bar{c}]\!]_{\mathcal{A}} \in [\![p]\!]_{\mathcal{A}}$

$\qquad$ iff (for all $a \in A$, $a \in [\![p]\!]_{\mathcal{A}}$) implies $[\![\bar{c}]\!]_{\mathcal{A}} \in [\![p]\!]_{\mathcal{A}}$,

which is clearly true, since $[\![\bar{c}]\!]_{\mathcal{A}} \in A$. Since $\mathcal{A}$ was chosen arbitrarily, this shows that $\vDash \forall x p(x) \to p(\bar{c})$.

Now let's unravel the second sentence:

$\mathcal{A} \vDash \forall x(p(x) \to p(\bar{c}))$ iff for all $a \in A$, $\mathcal{A} \vDash p(\tilde{a}) \to p(\bar{c})$

$\qquad$ iff for all $a \in A$, if $\mathcal{A} \vDash p(\tilde{a})$ then $\mathcal{A} \vDash p(\bar{c})$[61]

$\qquad$ iff for all $a \in A$, if $[\![\tilde{a}]\!]_{\mathcal{A}} \in [\![p]\!]_{\mathcal{A}}$, then $[\![\bar{c}]\!]_{\mathcal{A}} \in [\![p]\!]_{\mathcal{A}}$

$\qquad$ iff for all $a \in A$, if $a \in [\![p]\!]_{\mathcal{A}}$ then $[\![\bar{c}]\!]_{\mathcal{A}} \in [\![p]\!]_{\mathcal{A}}$.

But this certainly does not hold in all models. For example, if $A = \mathbb{N}$, $[\![p]\!]_{\mathcal{A}}$ is the set of all prime numbers, and $[\![\bar{c}]\!]_{\mathcal{A}} = 6$, then the statement becomes: "for all $a \in \mathbb{N}$, if a is prime, then 6 is prime", which is clearly false (it fails, e.g., when $a = 2$). Thus $\nvDash \forall x(p(x) \to p(\bar{c}))$. $\qquad\square$

---

[58] Just like in propositional logic: whether a primitive proposition is true depends on the valuation.

[59] Remember that like negation, we read quantifiers as taking *narrow scope*, so this formula should be parsed as $(\forall x p(x)) \to p(\bar{c})$ as opposed to $\forall x(p(x) \to p(\bar{c}))$. In fact, while the former is valid, the latter is definitely not, as we will show.

[60] Here we are using "implies" to refer specifically to *material implication*, as per the definition of truth for the "$\to$" connective. In other words, either the antecedent is false, or the consequent is true.

[61] As above, and throughout, this is the "if...then" of material implication.

Exercise 41

Let $S = (p,q,r)$, where p and q are relation symbols of arity 1 and r is a relation symbol of arity 2. For each of the following, determine whether it is valid and justify your answer.

(a) $\forall x r(x,x) \rightarrow \exists y \forall x r(x,y)$

(b) $\forall x \exists y r(x,y) \rightarrow \exists y \forall x r(x,y)$

(c) $\exists y \forall x r(x,y) \rightarrow \forall x \exists y r(x,y)$

(d) $\forall x \forall y r(x,y) \rightarrow \forall z r(z,z)$

(e) $\forall x \forall y r(x,y) \leftrightarrow \forall y \forall x r(x,y)$

(f) $\exists x \exists y r(x,y) \leftrightarrow \exists y \exists x r(x,y)$

(g) $\forall x p(x) \rightarrow \forall y p(y)$

(h) $\exists x p(x) \rightarrow \exists y p(y)$

(i) $\forall x (p(x) \wedge q(x)) \rightarrow (\forall x p(x) \wedge \forall x q(x))$

(j) $(\exists x p(x) \wedge \exists x q(x)) \rightarrow \exists x (p(x) \wedge q(x))$

(k) $\forall y (\forall x p(x) \rightarrow p(y))$

(l) $\forall y (p(y) \rightarrow \forall x p(x))$

(m) $\exists x p(x) \leftrightarrow \neg \forall x \neg p(x)$

(n) $\forall x p(x) \leftrightarrow \neg \exists x \neg p(x)$

Solution: page 210

---

[62] This accords with a common implicit convention in mathematics; for example, when we say something like "if x is non-zero then x^2 is positive", what we mean (at least implicitly) is that this statement holds *for all* numbers x.

[63] Technically we need to specify the order that these variables appear in; for concreteness we may assume that they are presented in the same order in which they are listed in the set VAR. It turns out, however, that the order doesn't matter: see Exercise 43.

It will be useful to extend this notion of truth to include formulas that are not closed. The best way to do this, roughly speaking, is to interpret free variables as if they were universally quantified.[62] More precisely, given any formula φ with free variables $FV(\varphi) = \{y_1, \ldots, y_n\}$, the **closure of** φ is defined to be the formula[63]

$$cl(\varphi) = \forall y_1 \cdots \forall y_n \varphi,$$

and we write $\mathcal{A} \vDash \varphi$ to mean $\mathcal{A} \vDash cl(\varphi)$, and $\vDash \varphi$ to mean $\vDash cl(\varphi)$.

Thus, for example, with a bit of extra bookkeeping we can generalize Exercise 41(n) to apply to any formula φ:

Proposition 19 *Let S be a signature and let φ be any formula in the language generated by S. Then $\vDash \forall x \varphi \leftrightarrow \neg \exists x \neg \varphi$.*

Proof We need to show that given any model $\mathcal{A}$ for $\mathcal{S}$, we have $\mathcal{A} \vDash \forall x \varphi \leftrightarrow \neg \exists x \neg \varphi$. But we also need to take into account that φ may not be a sentence; it may contain many free variables aside from x.[64] So suppose that $FV(\forall x \varphi \leftrightarrow \neg \exists x \neg \varphi) = \{y_1, \ldots, y_n\}$; we must show that

[64] In fact, it may not contain x at all!

$$\mathcal{A} \vDash \forall y_1 \cdots \forall y_n (\forall x \varphi \leftrightarrow \neg \exists x \neg \varphi).$$

The key point here is that before we even assess the biconditional, we have to deal with the (implicit) universal quantification of the free variables. For this, of course, we follow the recursive definition of truth,[65] which tells us that the above is equivalent to:

[65] And the recursive definition of substitution!

for all $a_1, \ldots, a_n \in A$, $\mathcal{A} \vDash (\forall x \varphi \leftrightarrow \neg \exists x \neg \varphi)[\tilde{a}_1/y_1] \cdots [\tilde{a}_n/y_n]$.

Notation is getting a bit cumbersome here, so we adopt the "trick" of writing $\varphi(y_1, \ldots, y_n, x)$ to "mark" the variables that we are paying special attention to,[66] so the above can be rewritten as:

for all $a_1, \ldots, a_n \in A$, $\mathcal{A} \vDash \forall x \varphi(\tilde{a}_1, \ldots, \tilde{a}_n, x)$

$\leftrightarrow \neg \exists x \neg \varphi(\tilde{a}_1, \ldots, \tilde{a}_n, x)$.

[66] These variables include all the free variables we added universal quantifiers for, and also, of course, the variable x, which isn't free but is explicitly quantified in the original formula, and as such we will be paying special attention to it momentarily.

So now our job is to evaluate whether, given arbitrary elements $a_1, \ldots, a_n$ in the domain of discourse, $\mathcal{A}$ satisfies

$$\forall x \varphi(\tilde{a}_1, \ldots, \tilde{a}_n, x) \leftrightarrow \neg \exists x \neg \varphi(\tilde{a}_1, \ldots, \tilde{a}_n, x).$$

In other words, we have reduced the question to one in which the formula we need to evaluate is a sentence![67] So we can proceed essentially as in the case of Exercise 41(n).

[67] Because each and every free variable has been replaced by a constant symbol.

Fix $a_1, \ldots, a_n \in A$, and for notational convenience, write $\hat{\varphi}(x)$ as an abbreviation for $\varphi(\tilde{a}_1, \ldots, \tilde{a}_n, x)$. Then we know that

$$\mathcal{A} \vDash \forall x \hat{\varphi}(x) \leftrightarrow \neg \exists x \neg \hat{\varphi}(x)$$

just in case

$$\mathcal{A} \vDash \forall x \hat{\varphi}(x) \text{ if and only if } \mathcal{A} \vDash \neg \exists x \neg \hat{\varphi}(x).$$

Let's start with the more complicated side, the right-hand side. We have:[68]

$$\mathcal{A} \vDash \neg \exists x \neg \hat{\varphi}(x) \text{ iff } \mathcal{A} \nvDash \exists x \neg \hat{\varphi}(x)$$

$$\text{iff it is not the case that for some } a{\in}A, \mathcal{A} \vDash \neg \hat{\varphi}(\tilde{a})$$

$$\text{iff it is not the case that for some } a{\in}A, \mathcal{A} \nvDash \hat{\varphi}(\tilde{a})$$

$$\text{iff for every } a \in A, \mathcal{A} \vDash \hat{\varphi}(\tilde{a})$$

$$\text{iff } \mathcal{A} \vDash \forall x \hat{\varphi}(x),$$

as desired.   □

In light of this, we can see now that we did not really need to include both universal and existential quantification in our language—Proposition 19 shows that we could've defined $\forall x$ as an abbreviation of $\neg \exists x \neg$; similarly, it easily follows that we could've also defined $\exists x$ as an abbreviation of $\neg \forall x \neg$.[69] Henceforth, we treat $\exists x$ as an abbreviation exactly in this way.

[68] You should check to make sure the following equivalences hold true even when x does not occur free in $\hat{\varphi}$, in which case $\hat{\varphi}(\tilde{a})$ is just $\hat{\varphi}$.

[69] This relationship between the two quantifiers is sometimes described by saying they are *dual* to one another.

Exercise 42

There is a similar sense in which conjunction and disjunction are "dual". What is it? And why might we expect the same theme to show up in the relationship between conjunction and disjunction as shows up in the relationship between universal and existential quantification?

Exercise 43

Let $\mathcal{S}$ be a signature and let φ be any formula in the language generated by $\mathcal{S}$.

(a) Show that $\vDash \forall x \forall y \varphi \leftrightarrow \forall y \forall x \varphi$.
(b) More generally, show that for any $y_1, \ldots, y_n \in$ VAR, we have

$$\vDash \forall y_1 \forall y_2 \cdots \forall y_n \varphi \leftrightarrow \forall y_{\pi(1)} \forall y_{\pi(2)} \cdots \forall y_{\pi(n)} \varphi,$$

where $\pi \colon \{1, \ldots, n\} \to \{1, \ldots, n\}$ is any bijection.

To close this section, we observe that every propositional tautology is valid. This is perhaps not terribly surprising, and the proof is not terribly profound—after all, we built predicate logic to be an extension of propositional logic, and gave all the propositional connectives essentially the same recursive truth conditions. However, to even state the result formally, we must be precise about what exactly a "tautology" is in the language of predicate logic.

Exercise 44

Let γ be a formula of propositional logic and let $p_1,\ldots,p_n$ denote the primitive propositions it contains; let $\varphi_1,\ldots,\varphi_n$ be any formulas of first-order logic.

(a) Recursively define (for all γ) the *substitution* of $\varphi_1,\ldots,\varphi_n$ for $p_1,\ldots,p_n$ in γ, denoted

$$\gamma[\varphi_1/p_1]\cdots[\varphi_n/p_n].$$

(b) Prove that if γ is a tautology of propositional logic, then

$$\vDash \gamma[\varphi_1/p_1]\cdots[\varphi_n/p_n].$$

2.3 Deduction

Deductions in predicate logic work in just the same way that deductions in propositional logic worked. As in the propositional case, there are many candidate axioms to choose from; for our purposes, the *axioms of predicate logic* consist of all instances of the following schemes:

(L1) $\varphi \to (\psi \to \varphi)$

(L2) $(\varphi \to (\psi \to \chi)) \to ((\varphi \to \psi) \to (\varphi \to \chi))$

(L3) $(\neg\psi \to \neg\varphi) \to (\varphi \to \psi)$

(L4) $\forall x \varphi(x) \to \varphi(t)$, provided t is free for x in φ[70]

(L5) $\forall x(\varphi \to \psi) \to (\varphi \to \forall x \psi)$, provided $x \notin FV(\varphi)$

(E1) $x = x$

(E2) $x_i = y \to f(x_1,\ldots,x_i,\ldots,x_n) = f(x_1,\ldots,y,\ldots,x_n)$, where f is any function symbol and $1 \le i \le n$

[70] A term t is said to be *free for x in φ* if no variable in t becomes bound in $\varphi[t/x]$.

(E3) $x_i = y \rightarrow (r(x_1, \ldots, x_i, \ldots, x_n) \rightarrow r(x_1, \ldots, y, \ldots, x_n))$, where r is any relation symbol or equality, and $1 \leq i \leq n$.

The *rules of inference of predicate logic* include the familiar *modus ponens*[71] and a new rule called *generalization*: from φ infer $\forall x \varphi$.[72] Formally, given a set of formulas Γ, a **deduction from** Γ in predicate logic is a finite sequence $\varphi_1, \ldots, \varphi_n$ such that for each $1 \leq i \leq n$, either:

(1) φ_i is an axiom,[73]

(2) $\varphi_i \in \Gamma$,

(3) there exist $j, k < i$ such that φ_k is $\varphi_j \rightarrow \varphi_i$, or

(4) there exists $j < i$ and $x \in$ VAR such that φ_i is $\forall x \varphi_j$.

Such a sequence is said to be a *deduction of φ_n from* Γ; we write $\Gamma \vdash \varphi$ to indicate that there exists a deduction of φ from Γ, and $\vdash \varphi$ as a shorthand for $\varnothing \vdash \varphi$.

The axioms (L1)–(L3), together with modus ponens, are exactly the same as in propositional logic, making the current deductive system an extension of that one. (L4) and (L5), together with generalization, govern quantification,[74] while (E1)–(E3) are included because the equality relation has special properties that are true in any model, which we therefore want to capture axiomatically. It will be helpful to dive a bit deeper into the intuitions behind these new additions to the deductive system before pressing on.

The axioms for equality are fairly straightforward, particularly (E1), which should certainly hold. (E2) and (E3) are essentially saying that functions and relations should treat equals identically—intuitively, if $x = y$, then if you substitute y for x inside a function or relation, it shouldn't change the value of the function or the truth of the relation.[75] Note that (E3) applies not only to relations in the signature, but also to equality itself; using this, one can show that additional axioms asserting that $=$ ought to be an equivalence relation are unnecessary, since they can be proved in the system.

---

[71] From φ and $\varphi \rightarrow \psi$ infer ψ.

[72] Abbreviated "Gen".

[73] That is, an instance of (L1)–(L5) or (E1)–(E3).

[74] We don't need any specific axioms for existential quantification because we are now treating it as an abbreviation, as discussed.

[75] A fancy way of saying this is that all functions and relations are *well-defined* with respect to the equivalence relation given by =. To get a sense of what this might mean, take a look at Exercise 46.

Exercise 45

Prove the following by writing out completely formal deductions.

(a) $\vdash \forall x(x = x)$
(b) $\vdash \forall x \forall y(x = y \rightarrow y = x)$
(c) $\vdash \forall x \forall y \forall z(x = y \rightarrow (y = z \rightarrow x = z))$

Solution: page 212

Exercise 46

Let $S = (a)$, where a is a function symbol of arity 2, and consider the S-structure $\mathcal{A} = (\mathbb{Z}, +)$ (so $[\![a]\!]_{\mathcal{A}} = +$). Instead of interpreting the equality symbol $=$ as true equality, suppose instead we use the following recursive clause:

$$[\![t_1 = t_2]\!]_{\mathcal{A}} = \textit{true} \text{ iff } [\![t_1]\!]_{\mathcal{A}} - [\![t_2]\!]_{\mathcal{A}} \text{ is even.}$$

(a) Show that all instances of the axioms (E1)–(E3) are still true in $\mathcal{A}$.
(b) Let $S' = (a, f)$, where a is as above and f is a function symbol of arity 1. Define a function $F \colon \mathbb{Z} \rightarrow \mathbb{Z}$ such that in the S'-structure $\mathcal{A}' = (\mathbb{Z}, +, F)$, with the equality symbol interpreted as in part (a), at least one instance of (E2) is no longer true.

Solution: page 213

Next let's consider (L4). The basic idea here is fairly easy to grasp, though the details take some work: $\forall x \varphi(x)$ means that φ is true for *all* x, so surely φ should be true whenever any particular thing is substituted for x.[76]

It's worth noting up front that nothing about this axiom implies that x must occur free (or even occur at all) in φ, but when it doesn't it's not hard to see that $\forall x \varphi$, φ, and $\varphi[t/x]$ are all equivalent, so (L4) certainly holds.

[76] Recall that $\varphi(t)$ is just a convenient notation for $\varphi[t/x]$.

> ### Exercise 47
>
> Let φ be a formula such that $x \notin FV(\varphi)$. Then $\vDash \forall x \varphi \leftrightarrow \varphi$ and $\vDash \varphi \leftrightarrow \varphi[t/x]$.

What about the condition that t be "free for x in φ"? The issue is that substitution of terms for variables, as we have defined it, technically allows for something rather strange to happen: it permits a variable in the term to become bound after the substitution. This can lead to strange results.

For instance, consider the formula $\forall x \exists y \ell(x, y)$. When we interpret this, say, over the domain of natural numbers, with ℓ interpreted as $<$, we get a true statement: for every natural number there is one that is greater. Since this is true for *every* number, it should remain true no matter what we substitute for x—this is the intuition of (L4). This would be written

$$\forall x \exists y \ell(x, y) \rightarrow \exists y \ell(t, y),$$

and it does *look* correct—whatever t is, there must be something greater than it. Right?

But consider a case where t includes the variable y; for instance, suppose t is $s(y)$ where s is interpreted as the successor function. Now, when t is substituted for x, the variable y becomes bound,[77] and we obtain

[77] That is, $s(y)$ is not free for x in $\exists y \ell(x, y)$.

$$\forall x \exists y \ell(x, y) \rightarrow \exists y \ell(s(y), y),$$

the consequent of which means something quite different from what we might have expected! It says that there exists a number that is greater than its successor, which is clearly false in the model given above.[78] Examples like this demonstrate why the restriction in (L4) to terms that are free for the quantifed variable is important.

[78] Choosing t to be y would give a similar problem: the consequent would assert the existence of a number greater than itself.

Turning to (L5), the intuition isn't too complicated: if $x \notin FV(\varphi)$, then in a sense the quantifier $\forall x$ "ignores" that part of the formula.[79] Consider, for example, the following informal statement: "for all people x, if the coffee has run out, then x is sad". Notice this has the rough form of the antecedent

[79] We have already seen a version of this cashed out semantically in Exercise 47; see also Proposition 21.

of (L5)—it is a universally quantified implication whose antecedent doesn't actually contain the variable being quantified. It shouldn't be too hard to convince yourself that this statement implies "if the coffee has run out, then everyone is sad". In other words, we can "move" the universal quantifier past the antecedent and into the consequent. This is (L5) in action. As we will see, this axiom plays a key role in our proof of the Deduction Theorem for first-order logic (Theorem 20).

Finally we come to the new rule of inference, generalization, which often causes some confusion. At first glance it may appear to be reversing the direction of implication in (L4), which seems wrong! Intuitively, just because something happens to be true for some particular x, that doesn't mean it's true for all x. And indeed, formulas of the form $\varphi \rightarrow \forall x \varphi$ are *not* valid in general.[80]

How can generalization be a sound rule of inference, then? The key insight here is that there's a big difference between asserting that the implication $\varphi \rightarrow \forall x \varphi$ is deducible (it's not!), versus asserting that *if* φ is deducible, then so is $\forall x \varphi$, which is what generalization says.[81] The assumption that φ is deducible is a strong one: since none of our axioms prioritize one variable over any other, if we can deduce that φ is true and φ contains x as a free variable, then this proof should work for any "arbitrary" x, so intuitively it ought to follow that the statement holds *for all* x.

We can see the analogous distinction at play in the semantic realm: there's a big difference between asserting that $\varphi \rightarrow \forall x \varphi$ is valid (it's not!), versus asserting that *if* φ is valid, then so is $\forall x \varphi$, which essentially follows from the definition of validity for formulas with free variables.[82] Preserving validity is exactly what we want rules of inference to do in order to prove soundness. So from a purely pragmatic perspective, this is another reason to accept—or at least not object to—generalization as a rule of inference: it preserves validity, so it won't be an obstacle to proving soundness.

[80] It may be illuminating here to compare Exercise 41(k), which is a universally quantified instance of (L4), with Exercise 41(l), which reverses the implication. The former is valid, while the latter is not.

[81] Since the Deduction Theorem is about importing assumptions as antecedents to conditionals, we'll see that this same distinction becomes relevant there, too.

[82] See Proposition 23.

As mentioned, the role of generalization becomes particularly relevant in the proof of the analogue of the Deduction Theorem for predicate logic. Because of the way we have set up the definition of deduction, we can see that whenever $\varphi(x) \in \Gamma$, we must have $\Gamma \vdash \forall x \varphi(x)$. Thus, if the analogue of the Deduction Theorem were to hold in predicate logic without restriction, we could reason from the fact that $\{\varphi(x)\} \vdash \forall x \varphi(x)$ to conclude that $\vdash \varphi(x) \rightarrow \forall x \varphi(x)$, which is not a valid formula.[83] So if we have any hope of proving soundness of our deductive system, it had better not be the case that the Deduction Theorem holds without restriction! And indeed, it does not (since we *can* prove soundness); the appropriately restricted form of the Deduction Theorem is the following.

[83] Again, compare this to Exercise 41(l).

Theorem 20 (The Deduction Theorem (for predicate logic))
Let φ and ψ be formulas and Γ a set of formulas in some first-order language. Then if $\Gamma \cup \{\varphi\} \vdash \psi$ via a deduction that does not employ the generalization rule of inference for any free variable of φ, then $\Gamma \vdash \varphi \rightarrow \psi$. Conversely, if $\Gamma \vdash \varphi \rightarrow \psi$, then $\Gamma \cup \{\varphi\} \vdash \psi$.

Proof One implication is exactly the same argument as in the Deduction Theorem for propositional logic (Theorem 8), so we focus on the other.

Assume that $\Gamma \cup \{\varphi\} \vdash \psi$ via a deduction that does not use generalization on any free variable of φ. We proceed by induction on the length n of this deduction of ψ. The base case, when $n = 1$, is handled in exactly the same way as the base case for the Deduction Theorem from propositional logic. Supposing inductively that the result is true for all deductions of length $k \leq n$, we consider a deduction $\varphi_1, \ldots, \varphi_{n+1}$ of ψ from $\Gamma \cup \{\varphi\}$ in which no applications of generalization involve any free variables of φ. If φ_{n+1} is an axiom, or a member of Γ, or is equal to φ, then the result follows as in the base case. And if φ_{n+1} follows from two previous lines by modus ponens, then again the same argument as in the proof of the propositional version works.

So all we must show is that the result holds when φ_{n+1} is $\forall x \varphi_j$ for some $x \in \text{VAR}$ and $j \leq n$. In this case, by assumption we know that $x \notin FV(\varphi)$. Moreover, by the inductive hypothesis, we have $\Gamma \vdash \varphi \rightarrow \varphi_j$. Thus, by generalization we know that $\Gamma \vdash \forall x(\varphi \rightarrow \varphi_j)$. Now since $x \notin FV(\varphi)$, using (L5) we can easily derive $\Gamma \vdash \varphi \rightarrow \forall x \varphi_j$, as desired.   □

Exercise 48

Suppose that x occurs free in $\varphi(x)$ and y does not occur in $\varphi(x)$. Show that $\vdash \forall x \varphi(x) \leftrightarrow \forall y \varphi(y)$.

Solution: page 214

2.4 Soundness and completeness

The path to proving soundness of the deductive system we have defined for predicate logic is very similar to that for propositional logic. We want to show that for every formula φ, if it is deducible, then it is valid. So we proceed by induction on the length of the deduction of φ: the base case then corresponds to checking that the axioms are all valid, and the inductive steps require us to show that the rules of inference preserve validity.

With the exception of (L4), which turns out to be pretty picky, proving all of this is relatively straightforward. The axioms (L1)–(L3) are valid in virtue of instantiating propositional tautologies (see Exercise 44). Establishing the validity of (L5) is a straightforward exercise.

Proposition 21 *Let φ be any formula with $x \notin FV(\varphi)$. Then*

$$\vDash \forall x(\varphi \rightarrow \psi) \rightarrow (\varphi \rightarrow \forall x \psi).$$

Proof Let $\mathcal{A}$ be a structure appropriate for the background signature, and let $y_1, \ldots, y_n$ denote the free variables of $\forall x(\varphi \rightarrow \psi) \rightarrow (\varphi \rightarrow \forall x \psi)$. Then we need to show that:[84]

$$\mathcal{A} \vDash \forall y_1 \cdots \forall y_n (\forall x(\varphi \rightarrow \psi) \rightarrow (\varphi \rightarrow \forall x \psi)).$$

[84] By definition of validity for formulas with free variables.

Thus, writing $\varphi(y_1, \ldots, y_n)$ for φ and $\psi(x, y_1, \ldots, y_n)$ for ψ, we need to show that for each n-tuple $(a_1, \ldots, a_n) \in A^n$, $\mathcal{A}$ satisfies

$$\forall x(\varphi(\tilde{a}_1, \ldots, \tilde{a}_n) \rightarrow \psi(x, \tilde{a}_1, \ldots, \tilde{a}_n))$$
$$\rightarrow (\varphi(\tilde{a}_1, \ldots, \tilde{a}_n) \rightarrow \forall x \psi(x, \tilde{a}_1, \ldots, \tilde{a}_n)).$$

For convenience, we will write $\hat{\varphi}$ for $\varphi(\tilde{a}_1, \ldots, \tilde{a}_n)$ and $\hat{\psi}(x)$ for $\psi(x, \tilde{a}_1, \ldots, \tilde{a}_n)$.[85]

[85] So our task is to show that
$$\mathcal{A} \vDash \forall x(\hat{\varphi} \rightarrow \hat{\psi}(x))$$
$$\rightarrow (\hat{\varphi} \rightarrow \forall x \hat{\psi}(x)).$$

Although this looks a lot like what we originally started with (except for the hats), for each previously free variable y_i in the original formula, we have now substituted some arbitrary constant $\tilde{a}_i$. This produces a closed formula so we can apply the recursive definition of truth. Note that this first part of the proof heavily parallels the first part of the proof of Proposition 19: in both cases we are simply replacing each free variable with some arbitrary constant to render the formula closed. Nothing terribly deep is happening here.

So suppose that $\mathcal{A} \vDash \forall x(\hat{\varphi} \rightarrow \hat{\psi}(x))$ and $\mathcal{A} \vDash \hat{\varphi}$; we wish to show that $\mathcal{A} \vDash \forall x \hat{\psi}(x)$. From $\mathcal{A} \vDash \forall x(\hat{\varphi} \rightarrow \hat{\psi}(x))$ we know that for all $a \in A$, $\mathcal{A} \vDash \hat{\varphi} \rightarrow \hat{\psi}(\tilde{a})$ (since x is not free in $\hat{\varphi}$); therefore, since $\mathcal{A} \vDash \hat{\varphi}$, we can conclude that for all $a \in A$, $\mathcal{A} \vDash \hat{\psi}(\tilde{a})$, from which it follows that $\mathcal{A} \vDash \forall x \hat{\psi}(x)$, as desired. $\square$

The validity of (E1)–(E3) is also not difficult to prove; it follows fairly directly from the fact that the equality symbol is always interpreted as actual equality in any model.[86]

[86] As always, don't forget to account for the possibility of free variables, as demonstrated in Propositions 19 and 21.

Exercise 49

Prove that every instance of (E1)–(E3) is valid.

In fact, we can prove the following more versatile result, which will be very useful for us later.

Lemma 22 (The Substitution Lemma) *Let t and s be closed terms.*

(a) *For any term u, $\vDash t = s \rightarrow u[t/x] = u[s/x]$.*
(b) *For any formula φ, $\vDash t = s \rightarrow (\varphi[t/x] \leftrightarrow \varphi[s/x])$.*

Proof Let $\mathcal{A}$ be a structure appropriate for the background signature. For part (a), we want to show that

$$\mathcal{A} \vDash t = s \rightarrow u[t/x] = u[s/x].$$

As always, this means we must show that

$$\mathcal{A} \vDash \forall y_1 \cdots \forall y_n (t = s \rightarrow u[t/x] = u[s/x]),$$

where $y_1, \ldots, y_n$ are the free variables of the original formula. Since t is closed, substitution doesn't affect it, so it is easy to see that given any $a_1, \ldots, a_n \in A$, we have

$$u[t/x][\tilde{a}_1/y_1]\cdots[\tilde{a}_n/y_n] = \hat{u}[t/x],$$

adopting our usual convention of writing $\hat{u}$ for $u[\tilde{a}_1/y_1]\cdots$ $[\tilde{a}_n/y_n]$.[87] Similarly, since s is closed, we have

$$u[s/x][\tilde{a}_1/y_1]\cdots[\tilde{a}_n/y_n] = \hat{u}[s/x].$$

Thus, our goal is to show that for each $(a_1,\ldots,a_n)\in A^n$,

$$\mathcal{A}\vDash t = s \to \hat{u}[t/x] = \hat{u}[s/x].$$

We assume $\mathcal{A}\vDash t = s$ and seek to prove $\mathcal{A}\vDash \hat{u}[t/x] = \hat{u}[s/x]$. For this we use structural induction; in particular, we prove the following: *for all terms v which contain no variables except possibly x,* $\mathcal{A}\vDash v[t/x] = v[s/x]$.[88]

If v is a constant symbol or x itself, this is immediate, which takes care of the base cases. Now suppose inductively the result holds for $v_1,\ldots,v_k$; we wish to show it holds also for $f(v_1,\ldots,v_k)$. We have:

$$
\begin{aligned}
[\![f(v_1,\ldots,v_k)[t/x]]\!]_{\mathcal{A}} &= [\![f(v_1[t/x],\ldots,v_k[t/x])]\!]_{\mathcal{A}}\\
&= [\![f]\!]_{\mathcal{A}}([\![v_1[t/x]]\!]_{\mathcal{A}},\ldots,[\![v_k[t/x]]\!]_{\mathcal{A}})\\
&= [\![f]\!]_{\mathcal{A}}([\![v_1[s/x]]\!]_{\mathcal{A}},\ldots,[\![v_k[s/x]]\!]_{\mathcal{A}})\\
&= [\![f(v_1[s/x],\ldots,v_k[s/x])]\!]_{\mathcal{A}}\\
&= [\![f(v_1,\ldots,v_k)[s/x]]\!]_{\mathcal{A}},
\end{aligned}
$$

where the third equality follows from the inductive hypothesis, which guarantees that for each $1 \le i \le k$, $\mathcal{A}\vDash v_i[t/x] = v_i[s/x]$, and therefore $[\![v_i[t/x]]\!]_{\mathcal{A}} = [\![v_i[s/x]]\!]_{\mathcal{A}}$. This completes the induction and so finishes our proof of part (a).

For part (b), reasoning analogous to the above leads us to the goal of showing that for each $(a_1,\ldots,a_n)\in A^n$,

$$\mathcal{A}\vDash t = s \to (\hat{\varphi}[t/x] \leftrightarrow \hat{\varphi}[s/x]),$$

where $\hat{\varphi}$ stands for $\varphi[\tilde{a}_1/y_1]\cdots[\tilde{a}_n/y_n]$. Assume $\mathcal{A}\vDash t = s$; we must show that $\mathcal{A}\vDash \hat{\varphi}[t/x] \leftrightarrow \hat{\varphi}[s/x]$. As above, we use induction to prove the following: *for all formulas ψ which contain no free*

[87] Try to prove this! It's a good warmup for Lemma 24, which weakens the assumption that t be closed.

[88] Of course, $\hat{u}$ is a term that contains no variables except possibly x.

variables except possibly x, $\mathcal{A} \models \psi[t/x] \leftrightarrow \psi[s/x]$. This time, however, we induct on the total number of connectives and quantifiers in ψ.[89] Denote this number by $\#\psi$.[90]

In the base case $\#\psi = 0$, ψ has no connectives nor any quantifiers, so it must be either of the form $r(t_1, \ldots, t_k)$ or $t_1 = t_2$. In either case, it follows that each t_i is a term that contains no variables except possibly x, so by part (a) we know that $[\![t_i[t/x]]\!]_{\mathcal{A}} = [\![t_i[s/x]]\!]_{\mathcal{A}}$. Then, similar to the above, we have

$$\mathcal{A} \models r(t_1, \ldots, t_k)[t/x] \text{ iff } \mathcal{A} \models r(t_1[t/x], \ldots, t_k[t/x])$$
$$\text{iff } ([\![t_1[t/x]]\!]_{\mathcal{A}}, \ldots, [\![t_k[t/x]]\!]_{\mathcal{A}}) \in [\![r]\!]_{\mathcal{A}}$$
$$\text{iff } ([\![t_1[s/x]]\!]_{\mathcal{A}}, \ldots, [\![t_k[s/x]]\!]_{\mathcal{A}}) \in [\![r]\!]_{\mathcal{A}}$$
$$\text{iff } \mathcal{A} \models r(t_1[s/x], \ldots, t_k[s/x])$$
$$\text{iff } \mathcal{A} \models r(t_1, \ldots, t_k)[s/x],$$

and likewise

$$\mathcal{A} \models (t_1 = t_2)[t/x] \text{ iff } \mathcal{A} \models t_1[t/x] = t_2[t/x]$$
$$\text{iff } [\![t_1[t/x]]\!]_{\mathcal{A}} = [\![t_2[t/x]]\!]_{\mathcal{A}}$$
$$\text{iff } [\![t_1[s/x]]\!]_{\mathcal{A}} = [\![t_2[s/x]]\!]_{\mathcal{A}}$$
$$\text{iff } \mathcal{A} \models t_1[s/x] = t_2[s/x]$$
$$\text{iff } \mathcal{A} \models (t_1 = t_2)[s/x].$$

With the base case done, we turn to the inductive step. Suppose the result holds for all ψ with $\#\psi \leq m$; we will show it holds also when $\#\psi = m + 1$. There are three cases to consider: ψ is either of the form $\neg\psi'$, or $\psi_1 \rightarrow \psi_2$, or $\forall y \psi'$.

First suppose that ψ is $\neg\psi'$. Then we have

$$\mathcal{A} \models (\neg\psi')[t/x] \text{ iff } \mathcal{A} \models \neg(\psi'[t/x])$$
$$\text{iff } \mathcal{A} \models \neg(\psi'[s/x])$$
$$\text{iff } \mathcal{A} \models (\neg\psi')[s/x],$$

where the second equivalence follows from the inductive hypothesis, since $\#\psi' = m$.

---

[89] The reason we do it this way rather than by structural induction will become clear shortly: see sidenote 91.

[90] Of course, one can easily provide a recursive definition of $\#\varphi$.

Now suppose that ψ is $\psi_1 \rightarrow \psi_2$. Then

$$\mathcal{A} \vDash (\psi_1 \rightarrow \psi_2)[t/x] \text{ iff } \mathcal{A} \vDash \psi_1[t/x] \rightarrow \psi_2[t/x]$$
$$\text{iff } \mathcal{A} \vDash \psi_1[s/x] \rightarrow \psi_2[s/x]$$
$$\text{iff } \mathcal{A} \vDash (\psi_1 \rightarrow \psi_2)[s/x],$$

where as above the second equivalence follows from the inductive hypothesis, in this case because both $\#\psi_1 \leq m$ and $\#\psi_2 \leq m$.

Finally, suppose ψ is $\forall y \psi'$ for some $y \in \text{VAR}$. If y and x are distinct variables, then

$$\mathcal{A} \vDash (\forall y \psi')[t/x] \text{ iff } \mathcal{A} \vDash \forall y(\psi'[t/x])$$
$$\text{iff for all } a \in A, \mathcal{A} \vDash \psi'[t/x][\tilde{a}/y]$$
$$\text{iff for all } a \in A, \mathcal{A} \vDash \psi'[\tilde{a}/y][t/x]$$
$$\text{iff for all } a \in A, \mathcal{A} \vDash \psi'[\tilde{a}/y][s/x]$$
$$\text{iff for all } a \in A, \mathcal{A} \vDash \psi'[s/x][\tilde{a}/y]$$
$$\text{iff } \mathcal{A} \vDash \forall y(\psi'[s/x])$$
$$\text{iff } \mathcal{A} \vDash (\forall y \psi')[s/x],$$

where the third and fifth equivalence follow because t and s are closed (as in sidenote 87), while the fourth equivalence follows from the inductive hypothesis, since $\#(\psi'[\tilde{a}/y]) = m$.[91] On the other hand, if y is x, then substitution for x is effectively blocked, so

$$\mathcal{A} \vDash (\forall y \psi')[t/x] \text{ iff } \mathcal{A} \vDash \forall y \psi'$$
$$\text{iff } \mathcal{A} \vDash (\forall y \psi')[s/x].$$

In either case we have established the desired equivalence, which completes the induction. □

Turning our attention now to the rules of inference, it is easy to see that modus ponens preserves validity for the same reason it preserved tautologies in classical propositional logic. Meanwhile generalization, as previously discussed, preserves validity because of how we defined validity of formulas containing free variables.

[91] This is why we couldn't use structural induction—notice that although $\psi'[\tilde{a}/y]$ has one fewer quantifier than ψ, it is not technically a subformula of ψ because of the substitution!

Proposition 23 *If* $\vDash \varphi$ *then also* $\vDash \forall x \varphi$.

Proof Let $y_1, \ldots, y_n$ denote the free variables of φ. First suppose that x is not free in φ; then as we have seen (Exercise 47), for any structure $\mathcal{A}$ appropriate for the background signature, $\mathcal{A} \vDash \forall x \varphi$ iff $\mathcal{A} \vDash \varphi$, from which it follows that

$$\mathcal{A} \vDash \forall x \varphi \text{ iff } \mathcal{A} \vDash \forall y_1 \cdots \forall y_n \forall x \varphi$$
$$\text{iff } \mathcal{A} \vDash \forall y_1 \cdots \forall y_n \varphi$$
$$\text{iff } \mathcal{A} \vDash \varphi.$$

Now suppose that x is free in φ; say x is y_i. Then by definition we have

$$\vDash \varphi \text{ iff } \vDash \forall y_1 \cdots \forall y_n \varphi,$$

and

$$\vDash \forall x \varphi \text{ iff } \vDash \forall y_i \varphi \text{ iff } \vDash \forall y_1 \cdots \forall y_{i-1} \forall y_{i+1} \cdots \forall y_n \forall y_i \varphi.$$

By Exercise 43, the order of these quantifiers is irrelevant, which gives us $\vDash \varphi$ iff $\vDash \forall x \varphi$, as desired. $\square$

Finally, we come to (L4). As promised, it is something of a quagmire. A preliminary lemma will be helpful.

Lemma 24 *Let* φ *be a formula, let* x *and* y *be distinct variables, and let* $\tilde{a}$ *be a constant symbol. Then if* t *is free for* x *in* φ, *we have*

$$\varphi[t/x][\tilde{a}/y] = \varphi[\tilde{a}/y][t[\tilde{a}/y]/x].$$

Exercise 50

Prove Lemma 24, then show that it generalizes to multiple substitutions, that is,

$$\varphi[t/x][\tilde{a}_1/y_1] \cdots [\tilde{a}_n/y_n]$$
$$= (\varphi[\tilde{a}_1/y_1] \cdots [\tilde{a}_n/y_n])[(t[\tilde{a}_1/y_1] \cdots [\tilde{a}_n/y_n])/x].$$

Proposition 25 *Let* φ *be any formula and let* t *be free for* x *in* φ. *Then*

$$\vDash \forall x \varphi(x) \rightarrow \varphi(t).$$

Proof Let $\mathcal{A}$ be a structure appropriate for the background signature; we wish to show that $\mathcal{A} \vDash \forall x \varphi(x) \rightarrow \varphi(t)$. The first impediment we face is that, as usual, to prove this we in fact need to show that

$$\mathcal{A} \vDash \forall y_1 \cdots \forall y_n (\forall x \varphi(x) \rightarrow \varphi(t)),$$

where $y_1, \ldots, y_n$ are the variables that occur free in $\forall x \varphi(x) \rightarrow \varphi(t)$. But this is not as straightforward as it has been previously; here it is relevant that t is free for x in φ, since this implies that when we substitute constants $\tilde{a}_1, \ldots, \tilde{a}_n$ for $y_1, \ldots, y_n$, the resulting formula is still of the form $\forall x \hat{\varphi}(x) \rightarrow \hat{\varphi}(\hat{t})$, where $\hat{\varphi}$ is $\varphi[\tilde{a}_1/y_1] \cdots [\tilde{a}_n/y_n]$, and $\hat{t}$ is $t[\tilde{a}_1/y_1] \cdots [\tilde{a}_n/y_n]$. This follows from Exercise 50.[92]

Exercise 51

Show that the above can fail when t is not free for x in φ: that is, give an example of a formula φ and a term t such that

$$(\forall x \varphi(x) \rightarrow \varphi(t))[\tilde{a}_1/y_1] \cdots [\tilde{a}_n/y_n]$$

is *not* of the form $\forall x \hat{\varphi}(x) \rightarrow \hat{\varphi}(\hat{t})$. (It may help to choose φ so that $n = 1$.)

So we must show that

$$\mathcal{A} \vDash \forall x \hat{\varphi}(x) \rightarrow \hat{\varphi}(\hat{t}),$$

where the formula on the right is a sentence. This means we must show that if for all $a \in A$, $\mathcal{A} \vDash \hat{\varphi}(\tilde{a})$, then $\mathcal{A} \vDash \hat{\varphi}(\hat{t})$. The key intuition here is the observation that $[\![\hat{t}]\!]_{\mathcal{A}} \in A$,[93] so when we quantify over all $a \in A$, this *includes* $[\![\hat{t}]\!]_{\mathcal{A}}$. In order to use this fact, we need to show that $\hat{\varphi}(\hat{t})$ and $\hat{\varphi}(\widetilde{[\![\hat{t}]\!]_{\mathcal{A}}})$ are equivalent.

Recall from Exercise 40 that we know $\mathcal{A} \vDash \hat{t} = [\![\hat{t}]\!]_{\mathcal{A}}$. The Substitution Lemma (Lemma 22) then yields

$$\mathcal{A} \vDash \hat{\varphi}(\hat{t}) \leftrightarrow \hat{\varphi}(\widetilde{[\![\hat{t}]\!]_{\mathcal{A}}}).$$

Thus, if $\mathcal{A} \vDash \hat{\varphi}(\tilde{a})$ for all $a \in A$, then in particular $\mathcal{A} \vDash \hat{\varphi}(\widetilde{[\![\hat{t}]\!]_{\mathcal{A}}})$, hence $\mathcal{A} \vDash \hat{\varphi}(\hat{t})$, as desired. $\quad\square$

---

[92] Or, in other words, repeated applications of Lemma 24.

[93] Since t is free for x in φ, none of the free variables of t become bound in $\varphi(t)$, from which it follows that $FV(t) \subseteq \{y_1, \ldots, y_n\}$, and therefore $\hat{t}$ must be closed as well.

Theorem 26 (Soundness) *If* $\vdash \varphi$ *then* $\vDash \varphi$.

Proof Induction on the length of the deduction of φ, analogous to Theorem 10. All the real work for this has been carried out above: the base case requires knowing that the axioms are all valid (Exercise 44, Proposition 25, Proposition 21, Exercise 49), and the inductive step requires knowing that the rules of inference preserve validity (Proposition 23). That's it!   □

Exercise 52

Consider the following argument: the sequence

φ_1: $\forall x \exists y r(x,y)$ (assump)
φ_2: $\forall x \exists y r(x,y) \to \exists y r(y,y)$ (L4)
φ_3: $\exists y r(y,y)$ (1,2 MP)

shows that $\{\forall x \exists y r(x,y)\} \vdash \exists y r(y,y)$; therefore, by the Deduction Theorem (Theorem 20), we have

$$\vdash \forall x \exists y r(x,y) \to \exists y r(y,y).$$

(a) Prove that, in fact, $\nvdash \forall x \exists y r(x,y) \to \exists y r(y,y)$.
(b) In light of (a), something must be wrong with the argument above. Explain.

Solution: page 215

Exercise 53

Suppose the formula $\forall x \varphi(x) \to \psi$ is closed (so, in particular, x does not occur free in ψ). Is it the case that

$$\{\forall x \varphi(x) \to \psi\} \vdash \forall x (\varphi(x) \to \psi)?$$

Justify your answer.

Solution: page 215

---

[94] In fact, the proof of completeness for first-order logic has a reasonable claim to being one of the hardest proofs one might encounter in an undergraduate curriculum.

As was the case for propositional logic, proving completeness is significantly harder.[94] We'll use almost all of the machinery we developed for that previous completeness proof (Theorem 15), and much more besides.

Let's begin with a rough, high-level overview of the plan. We want to show that every valid formula is deducible; for this it will be enough to show that every consistent set of sentences is satisfiable.[95] This is sufficient because, as before, we can show that if φ is not deducible then $\{\neg\varphi\}$ is consistent, and thus satisfiable, so φ must be false in some model—which yields the contrapositive of completeness.[96]

So the central task is to show that given a consistent set of sentences Γ, it is satisfiable, or in other words, that it has a model. This is the major difference from the propositional case—building a model for a set of first-order sentences is *much* more involved than constructing a valuation.[97] As we know, a first-order model consists in a domain of discourse over which quantification occurs, together with interpretations for all the function, predicate, and constant symbols. Where are we supposed to get this universe of objects and these interpretations?

The answer, more or less, is that we construct the domain of discourse from the closed terms of the language. This makes a certain sense: closed terms are supposed to denote objects, anyway, and we might expect the proof of completeness to involve "creating" semantic structures from the syntax, in some shape or form.[98]

The details present two problems we must overcome. First, our language has equality in it, but if each closed term is taken to represent a *distinct* object, we run into difficulties given the fact that distinct terms may nonetheless denote the same object. In other words, the universe may be *too populous*. The solution in this case is fairly straightforward: rather than taking our objects to be individual closed terms, we will instead identify elements of the domain of discourse with equivalence classes of such terms.[99]

The second (and more vexing) problem stems from the fact that we need the model we construct to actually make all the

[95] Where, analogously to the notion from propositional logic, a set of sentences Γ is *satisfiable* if there is a model that satisfies it, that is, a model in which every sentence of Γ is true.

[96] This also relies on transforming formulas into equivalent sentences by taking the closure.

[97] Recall that even finding a valuation was highly non-trivial in the proof of completeness for propositional logic—it required the definition of maximally consistent sets.

[98] As we did in Theorem 15—valuations were built from maximally consistent sets.

[99] Where the equivalence relation is determined, roughly speaking, by which equalities are deducible from Γ.

formulas in Γ true. Now for the most part this can be arranged by defining the interpretations of the functions, relations, and constants properly, using Γ, and by extending Γ to a maximally consistent set. This is similar to the work we did in the propositional case. But the interpretation of quantified formulas presents a unique challenge: if Γ contains a formula of the form $\exists x \varphi(x)$, then we need to be sure that our domain of discourse *actually includes* an object that witnesses this quantifier.[100] And, in general, it may not. So to get the correspondence to work out the way we need it to, we'll actually be adding new constant symbols to the language, whose sole purpose is to be potential witnesses for all the possible existential statements. This is a delicate process, in part because expanding the language to obtain witnesses to existentially quantified formulas also, simultaneously, *creates* new existentially quantified formulas! This is the task that we will begin the proof with.

To simplify the proof, we will restrict our attention to countable signatures.[101] It's not hard to show (similar to the propositional case) that the language generated by a countable signature is also countable.[102]

> ### Exercise 54
>
> Show that if S is a countable signature then the language generated by S is also countable.

As in the propositional case, we say that a set of sentences Γ is **consistent** if there is no formula ψ such that $\Gamma \vdash \psi$ and $\Gamma \vdash \neg\psi$.

Lemma 27 (Model Existence Lemma) *Let S be a countable signature and let Γ be a consistent set of sentences in the language generated by S. Then Γ is satisfiable.*

Proof As promised, we start by adding new constants to the language in an effort to ensure that all (true) existentially quantified statements will (eventually, in the model we construct) have witnesses. To this end, let $D = \{\bar{d}_1, \bar{d}_2, \ldots\}$ be a

[100] That is, there must actually be some object a in the domain such that $\varphi(a)$ is true.

[101] A signature is called countable if it includes only countably many symbols, which of course is the case precisely when the set of constant symbols is countable.

[102] That is, contains only countably many formulas.

countable collection of new[103] constant symbols. Let S' denote the signature obtained by adding these new constant symbols to S.

Since S' is countable, by Exercise 54 so is the language generated by S'; therefore, we can enumerate all the formulas in this language that contain exactly one free variable:

$$\varphi_1(y_1), \varphi_2(y_2), \ldots.$$

Choose $\bar{e}_1 \in D$ such that $\bar{e}_1$ does not occur in $\varphi_1(y_1)$; then recursively choose $\bar{e}_k \in D$ such that $\bar{e}_k \notin \{\bar{e}_1, \ldots, \bar{e}_{k-1}\}$ and $\bar{e}_k$ does not occur in any of $\varphi_1(y_1), \ldots, \varphi_k(y_k)$.[104,105]

For each k, let ψ_k denote the sentence

$$\exists y_k \varphi_k(y_k) \to \varphi_k(\bar{e}_k).$$

[104] Of course this is always possible since this is a finite list, and so can only contain finitely many of the constant symbols in D.

[105] The point of choosing the $\bar{e}_k$ in this careful way is to ensure that they are "free" to be used as witnesses, as we shall see.

Loosely speaking, ψ_k asserts that *if* there is something that satisfies $\varphi_k(y_k)$, then $\bar{e}_k$ does. These sentences will be crucial for ensuring that every (true) existentially quantified sentence has a witness (in the model we still have to construct). In order for the sentences ψ_k to actually play this role, however, we need to add them to Γ. And in order to add them to Γ without creating problems, we need to be sure we can do so consistently: that is, we need to show that $\Gamma \cup \{\psi_k\}_{k=1}^{\infty}$ is consistent.

To show this, set $\Gamma_0 = \Gamma$ and recursively define $\Gamma_k = \Gamma_{k-1} \cup \{\psi_k\}$. We will show that for each k, Γ_k is consistent, and use this to conclude that

$$\Gamma_\infty = \bigcup_{k=0}^{\infty} \Gamma_k = \Gamma \cup \{\psi_k\}_{k=1}^{\infty}$$

is consistent as well.

Naturally, to show that for all k, Γ_k is consistent, we use induction. The base case $k = 0$ looks obvious—after all, $\Gamma_0 = \Gamma$ and Γ is consistent by assumption. But actually there is a subtlety here. Our assumption that Γ is consistent was made in the *original* language generated by S. But for $k \geq 1$, the set Γ_k can only be understood as a set of sentences in the expanded language generated by S'; therefore, to be useful in the induction, we

[106] What's the difference? The expanded language has more axioms (because there are more instances of the axiom schemes)! When we need to emphasize this distinction, we will henceforth refer to a set being consistent *over* a signature.

must also show that Γ_0 is consistent *over S'*.[106] Fortunately, this isn't too hard.

Exercise 55

Show that if Γ is consistent over S and S' extends S with extra constant symbols, then Γ is also consistent over S'.

Hint: page 216

This takes care of the base case, so we turn next to the inductive step. Let $k > 0$ and suppose inductively that Γ_{k-1} is consistent; we wish to show that Γ_k is also consistent. Suppose for contradiction it is not. Then, by the principle of explosion,[107] we must have $\Gamma_k \vdash \neg\psi_k$. Thus, since $\Gamma_k = \Gamma_{k-1} \cup \{\psi_k\}$, we have

[107] The principle of explosion holds in predicate logic for exactly the same reason it held in propositional logic (Exercise 24).

$$\Gamma_{k-1} \cup \{\psi_k\} \vdash \neg\psi_k.$$

The Deduction Theorem for predicate logic (Theorem 20) applies here, since ψ_k is a sentence and thus has no free variables, so we obtain

$$\Gamma_{k-1} \vdash \psi_k \rightarrow \neg\psi_k.$$

[108] As it does in propositional logic (Proposition 9).

And from this it follows that $\Gamma_{k-1} \vdash \neg\psi_k$.[108] Thus we have

$$\Gamma_{k-1} \vdash \neg(\exists y_k \varphi_k(y_k) \rightarrow \varphi_k(\bar{e}_k)),$$

[109] Again, using basic propositional reasoning.

so it follows that[109]

$$\Gamma_{k-1} \vdash \exists y_k \varphi_k(y_k)$$

and

$$\Gamma_{k-1} \vdash \neg\varphi_k(\bar{e}_k).$$

[110] It certainly doesn't occur in Γ, and we specifically chose $\bar{e}_k$ so that it did not occur in any of $\psi_1, \ldots, \psi_{k-1}$.

[111] That is, a variable that does not occur elsewhere in the deduction.

Now, by construction, we know that the constant symbol $\bar{e}_k$ does not occur in any sentence in Γ_{k-1}.[110] One can then easily check that, if we consider the deduction of $\neg\varphi_k(\bar{e}_k)$ from Γ_{k-1}, and replace each occurrence of $\bar{e}_k$ with a "fresh" variable x,[111] then we produce a deduction of $\neg\varphi_k(x)$ from Γ_{k-1}. Indeed, this follows from the following exercise.

Exercise 56

Let Γ be a set of formulas and φ a formula; let x be a variable that does not occur in Γ or φ, and let $\bar{c}$ be a constant that does not occur in Γ. Then, if $\Gamma \vdash \varphi(\bar{c})$, also $\Gamma \vdash \varphi(x)$.

So we have $\Gamma_{k-1} \vdash \neg\varphi_k(x)$ and so, by generalization, $\Gamma_{k-1} \vdash \forall x \neg\varphi_k(x)$. From this it follows that $\Gamma_{k-1} \vdash \forall y_k \neg\varphi_k(y_k)$, using Exercise 48. Thus we obtain $\Gamma_{k-1} \vdash \neg\exists y_k \varphi_k(y_k)$, a contradiction. This completes the induction.

So we've shown that each Γ_k is consistent. As we've seen before, since $\Gamma_0 \subseteq \Gamma_1 \subseteq \Gamma_2 \subseteq \cdots$, it follows that $\Gamma_\infty = \bigcup_{k=0}^{\infty} \Gamma_k$ must also be consistent.[112]

The next step is to extend Γ_∞ to a *maximally consistent* set: a consistent set that cannot be extended to any strictly larger consistent set. This can be done exactly as it was in propositional logic—in other words, the proof of Lindenbaum's Lemma (Lemma 13) works equally well in the present context. So let $\Gamma' \supseteq \Gamma_\infty$ be a maximally consistent set.[113]

We are (finally!) in a position to define the model that will satisfy Γ. As discussed, the domain of discourse is obtained by taking equivalence classes of closed terms. Specifically, let T denote the set of all closed terms of the language generated by $\mathcal{S}'$. Next, define a binary relation $\sim$ on T by setting

$$t \sim t' \text{ iff } t = t' \in \Gamma'.\text{[114]}$$

In fact, $\sim$ is an equivalence relation—this follows easily from Exercise 45.[115] Define $A = T/\sim$; that is, let the domain of discourse be the quotient of T by $\sim$.[116] This effectively forces two elements to be equal (in the quotient) precisely when Γ' "thinks" they are equal,[117] which guarantees a crucial correspondence between which equalities the model makes true and which are in Γ'.

Now that we have the domain of discourse specified, we can define the interpretation of the symbols in the signature $\mathcal{S}'$.

[112] Briefly: if not, then since deducing a contradiction can only use finitely many assumptions, it follows that Γ_N must also be inconsistent, for sufficiently large N.

[113] Just as in the propositional case, we can show that for every sentence φ, either $\varphi \in \Gamma'$ or $\neg\varphi \in \Gamma'$.

[114] Equivalently (because Γ' is maximally consistent), we can define $t \sim t'$ iff $\Gamma' \vdash t = t'$.

[115] And some applications of (L4).

[116] For a refresher on equivalence relations and quotients, see Appendix A.2.

[117] Where, just as in the propositional case, we read $\varphi \in \Gamma'$ loosely as Γ' "thinking" that φ holds.

Given $[t_1], \ldots, [t_n] \in A$ and a function symbol f of arity n, define $[\![f]\!]_{\mathcal{A}} : A^n \to A$ by:

$$[\![f]\!]_{\mathcal{A}}([t_1], \ldots, [t_n]) = [f(t_1, \ldots, t_n)].$$

This should be read very carefully, since the construction is mixing syntax and semantics in a potentially confusing way: it says that the *actual* function (from A^n to A) we interpret f to be in the model works by mapping any n-tuple of equivalence classes of the form $([t_1], \ldots, [t_n])$ to the equivalence class given by $[f(t_1, \ldots, t_n)]$. In order for this to actually define a function,[118] we must check that the output of the function is not sensitive to the particular choices of representatives $t_1, \ldots, t_n$ for the equivalence classes. This follows from (E2).

[118] That is, for $[\![f]\!]_{\mathcal{A}}$ to be "well-defined on the quotient".

Exercise 57

Prove that $[\![f]\!]_{\mathcal{A}}$ is well-defined on the quotient $T/{\sim}$.

Solution: page 216

The interpretation of relation symbols is similar: given a relation symbol r of arity n, define

$$[\![r]\!]_{\mathcal{A}} = \{([t_1], \ldots, [t_n]) \in A^n \; : \; r(t_1, \ldots, t_n) \in \Gamma'\}.$$

Again, reading carefully, we have: the n-tuple $([t_1], \ldots, [t_n]) \in A^n$ satisfies the relation just in case the formula $r(t_1, \ldots, t_n)$ is in Γ'. And again, we need to check that this set is well-defined, which follows from (E3).

Exercise 58

Suppose that for all $1 \le i \le n$ we have $t'_i \sim t_i$. Then

$$r(t'_1, \ldots, t'_n) \in \Gamma' \text{ iff } r(t_1, \ldots, t_n) \in \Gamma'.$$

Finally, for any constant symbol $\bar{c}$ in $\mathcal{S}'$, set

$$[\![\bar{c}]\!]_{\mathcal{A}} = [c].$$

The notation makes this look almost trivial, but it is saying something meaningful: the interpretation of the constant symbol $\bar{c}$ in the model—whose domain of discourse, remember, consists of equivalence classes of closed terms—is just the equivalence class of that very constant symbol.[119]

Let $\mathcal{A} = (A, [\![\cdot]\!]_{\mathcal{A}})$. The last big task on the table is to show that this model $\mathcal{A}$ satisfies *exactly* the sentences that are in Γ'. This is the predicate logic version of the Truth Lemma. Before tackling it, it will be helpful to establish a preliminary result.

Lemma 28 *For all closed terms $t \in T$, we have $[\![t]\!]_{\mathcal{A}} = [t]$.*

Proof Induction on the structure of t. For the base case, suppose that t is $\bar{c}$. Then we have $[\![\bar{c}]\!]_{\mathcal{A}} = [\bar{c}]$ by definition of the model $\mathcal{A}$.

Now suppose inductively the result holds for closed terms $t_1, \ldots, t_n$ and consider a closed term of the form $f(t_1, \ldots, t_n)$. We have:

$$[\![f(t_1, \ldots, t_n)]\!]_{\mathcal{A}} = [\![f]\!]_{\mathcal{A}}([\![t_1]\!]_{\mathcal{A}}, \ldots, [\![t_n]\!]_{\mathcal{A}})$$
$$= [\![f]\!]_{\mathcal{A}}([t_1], \ldots, [t_n])$$
$$= [f(t_1, \ldots, t_n)],$$

where the first line follows from the recursive definition of semantic interpretation, the second line uses the inductive hypothesis, and the last line comes from the definition of $\mathcal{A}$. This completes the induction. □

Lemma 29 (Truth Lemma) *For all sentences φ of the language generated by S', $[\![\varphi]\!]_{\mathcal{A}} = true$ if and only if $\varphi \in \Gamma'$.*

Proof We proceed by induction on the total number of connectives and quantifiers in φ, as in the proof of Lemma 22. As before, we denote this number by $\#\varphi$.

For the base case, when $\#\varphi = 0$, there are two possibilities: either φ is of the form $t_1 = t_2$ for some (closed) terms t_1 and t_2, or else φ is of the form $r(t_1, \ldots, t_n)$ for some relation symbol r (of arity n) and closed terms $t_1, \ldots, t_n$.[120]

[119] In a moment we will show (Lemma 28) that this correspondence extends to all closed terms.

[120] The terms must be closed because otherwise φ would not be a sentence (since in this base case it contains no quantifiers).

For the first case, we need to show that

$$[\![t_1 = t_2]\!]_{\mathcal{A}} = \text{true} \text{ iff } t_1 = t_2 \in \Gamma'.$$

By the recursive definition of truth, the former statement is equivalent to $[\![t_1]\!]_{\mathcal{A}} = [\![t_2]\!]_{\mathcal{A}}$. By Lemma 28, this in turn is equivalent to $[t_1] = [t_2]$. And since these are equivalence classes, we know they are equal iff $t_1 \sim t_2$, which by definition means $t_1 = t_2 \in \Gamma'$, as desired.

For the second case, we need to show that

$$[\![r(t_1, \ldots, t_n)]\!]_{\mathcal{A}} = \text{true} \text{ iff } r(t_1, \ldots, t_n) \in \Gamma'.$$

By the recursive definition of truth, the former statement is equivalent to

$$([\![t_1]\!]_{\mathcal{A}}, \ldots, [\![t_n]\!]_{\mathcal{A}}) \in [\![r]\!]_{\mathcal{A}},$$

which by Lemma 28 is equivalent to

$$([t_1], \ldots, [t_n]) \in [\![r]\!]_{\mathcal{A}},$$

which holds if and only if $r(t_1, \ldots, t_n) \in \Gamma'$, by definition of $\mathcal{A}$. This completes the base case.

Now suppose inductively the result is true for all formulas with up to k connectives and quantifiers, and let $\#\varphi = k + 1$. There are three possibilities for φ in this case: either φ is of the form $\neg\psi$, or of the form $\psi \to \chi$, or of the form $\forall x\psi$. The first two cases are handled exactly as they were in the proof of the Truth Lemma for propositional logic (Lemma 14)—namely, using the fact that Γ' is maximally consistent. So we focus here on the new case, where φ is the quantified sentence $\forall x\psi$.

First consider the case where x does *not* occur free in ψ. In this case, since φ is closed, so must be ψ; moreover, it's easy to see that $[\![\psi]\!]_{\mathcal{A}} = [\![\forall x\psi]\!]_{\mathcal{A}}$,[121] and also $\psi \in \Gamma'$ iff $\forall x\psi \in \Gamma'$.[122] Thus we have:

$$\forall x\psi \in \Gamma' \text{ iff } \psi \in \Gamma'$$
$$\text{iff } [\![\psi]\!]_{\mathcal{A}} = \text{true}$$
$$\text{iff } [\![\forall x\psi]\!]_{\mathcal{A}} = \text{true},$$

[121] This follows from the fact that $\psi[\bar{a}/x]$ is identical to ψ; see Exercise 47.

[122] One direction uses the generalization rule of inference, while the other direction uses (L4).

where the second equivalence uses the inductive hypothesis (since clearly $\#\psi < \#\varphi$).

So assume that x *does* occur free in $\psi(x)$. First suppose that $[\![\forall x\psi(x)]\!]_A = true$. We must show that $\forall x\psi(x) \in \Gamma'$. Suppose for contradiction that $\forall x\psi(x) \notin \Gamma'$. Then by maximality we must have $\neg\forall x\psi(x) \in \Gamma'$, and so $\exists x\neg\psi(x) \in \Gamma'$. Now, since x occurs free in $\psi(x)$ but we know that $\forall x\psi(x)$ is a sentence, that tells us that x is the *only* free variable in $\neg\psi(x)$. Therefore this formula occurs in the enumeration we began this proof with: $\neg\psi(x)$ is $\varphi_i(y_i)$ for some i. Hence, by construction of Γ_∞ (which is contained in Γ'), we know that

$$\exists x\neg\psi(x) \to \neg\psi(\bar{e}_i) \in \Gamma'.$$

Putting this together with the fact that $\exists x\neg\psi(x) \in \Gamma'$, we obtain $\neg\psi(\bar{e}_i) \in \Gamma'$, so $\psi(\bar{e}_i) \notin \Gamma'$. Clearly, $\#\psi(\bar{e}_i) < \#\varphi$,[123] so by the inductive hypothesis we have $[\![\psi(\bar{e}_i)]\!]_A = false$. But by supposition we have $[\![\forall x\psi(x)]\!]_A = true$; from this it follows (using soundness of (L4)) that $[\![\psi(\bar{e}_i)]\!]_A = true$, a contradiction.

Conversely, suppose that $\forall x\psi(x) \in \Gamma'$, and assume for contradiction that $[\![\forall x\psi(x)]\!]_A = false$. By the recursive definition of truth, this means that for some $[t] \in A$, we have $[\![\psi([\tilde{t}])]\!]_A = false$. By Lemma 28, we know that $[t] = [\![t]\!]_A$, so $[\tilde{t}] = \widetilde{[\![t]\!]_A}$; therefore, since $A \vDash t = \widetilde{[\![t]\!]_A}$,[124] we also have $A \vDash t = [\tilde{t}]$. The Substitution Lemma (Lemma 22) then yields $A \vDash \psi(t) \leftrightarrow \psi([\tilde{t}])$, hence

$$[\![\psi(t)]\!]_A = [\![\psi([\tilde{t}])]\!]_A = false.$$

The inductive hypothesis can then by applied to tell us that $\psi(t) \notin \Gamma'$. However, by supposition, $\forall x\psi(x) \in \Gamma'$, which implies (using (L4)) that $\psi(t) \in \Gamma'$, a contradiction. This completes the induction. □

From the Truth Lemma it easily follows that A satisfies Γ with respect to the language generated by S: we know that $\Gamma \subseteq \Gamma'$, and we can interpret the language generated by S (rather than S') in A simply by restricting the interpretation $[\![\cdot]\!]_A$ defined above to S.[125] □

[123] But note that $\psi(\bar{e}_i)$ is *not* a subformula of φ, which is why we chose to do induction on the number of connectives and quantifiers rather than the structure of φ.

[124] Exercise 40.

[125] Of course, we keep the definition of A the same—we still want all the extra elements of the domain to *exist*, since we need witnesses for all our existential quantifiers! We just no longer need those extra elements to correspond to constant symbols in the language.

126 The proof is an easy
    generalization of the
    argument given in
    sidenote 122.

The Model Existence Lemma can be applied directly to produce a proof of completeness, using the following simple result:[126]

Exercise 59

Show that for all formulas φ, we have $\vdash \varphi$ if and only if $\vdash cl(\varphi)$.

Corollary 30 (Completeness) *If $\vDash \varphi$ then $\vdash \varphi$.*

Proof Suppose that $\vDash \varphi$. Then by definition $\vDash cl(\varphi)$ as well. Now suppose for contradiction that $\nvdash \varphi$. Then by Exercise 59 we know that $\nvdash cl(\varphi)$, so $\{\neg cl(\varphi)\}$ is consistent. By the Model Existence Lemma (Lemma 27), this set is satisfiable, which means there is some model $\mathcal{A}$ with $\mathcal{A} \nvDash cl(\varphi)$, a contradiction. □

Exercise 60

Let Γ be a set of sentences over a countable language. As in propositional logic, we say that Γ **entails** φ and write $\Gamma \vDash \varphi$ if every model that satisfies Γ also makes φ true. Prove the following.

(a) $\Gamma \vdash \varphi$ if and only if $\Gamma \vDash \varphi$.
(b) Γ is satisfiable if and only if Γ is consistent.
(c) Γ is satisfiable if and only if every finite subset of Γ is satisfiable.

Solution: page 217

---

3 Modal Logic

Modal logics are expansions of classical logic designed for reasoning about a richer variety of phenomena. While classical propositional logic can connect statements[1] together with simple logical relations,[2] it cannot capture more complex relationships between statements, such as

- "It's raining" and "Alice believes that it's raining";
- "The value of x is 3" and "The value of x will continue to be 3";
- "Bob is wearing a blue hat" and "If Bob were wearing a red hat, he would know it".

Though we will develop the core logical formalism in full generality, for examples and intuition we'll return again and again to *epistemic* scenarios.[3] It's helpful to begin with a toy example where reasoning of this sort is natural and illuminating.

[1] For example, "It's raining" or "John is wearing a blue hat" or "The value of x is 3".

[2] For example, "and", "or", "not", "if … then" (the material conditional), and "if and only if".

[3] That is, interpretations that involve reasoning about *knowledge*.

3.1 Epistemic reasoning

You may have heard this puzzle before (or some variant thereof): three people—Alice, Bob, and Carl—sit facing one another. On each of their heads a hat has been placed without them seeing it. All they were told as a group is that each of the hats is either red or blue.

Each player can see every hat but their own. As it happens, they are all wearing red hats. The following announcement is made to the group: "At least one of you is wearing a red hat."

The players are then asked whether they know the colour of their own hat. They all answer "No." The question is repeated. They all give the same answer as before, "No." Finally, the question is repeated a third time, and at this point each player correctly identifies their own hat as red. How did they figure it out?

Exercise 61

Before reading further, take a minute to try to puzzle this out!

Let's consider things from Alice's perspective, for concreteness. At the outset of the game, she sees a red hat on both Bob and Carl, and has no idea whether her own hat is red or blue. When no one can identify their own hat colour the first time the question is asked, she's not surprised. After all, the only way someone could deduce the colour of their own hat simply from the announcement "at least one hat is red" is if both the other hats they are seeing are blue. And Alice can plainly see that no one is seeing two blue hats.[4]

But Alice can take this reasoning a step farther. She can imagine what (say) Bob thinks. It's possible (for all she knows) that Bob is seeing a blue hat on her. In that case, since for all he knows his own hat might also be blue, Bob would consider it possible that Carl is the only one wearing a red hat.[5] But in this scenario, when Carl answers the first question negatively, Bob would be able to deduce that in fact his own hat must be red—because if Carl's hat was the only red one, he'd have figured that out from the initial announcement.

So the upshot is that Alice can reason that *if* her hat is blue, then Bob will learn that his hat is red after the first question. In this case, Bob would answer yes to the second question. Since he doesn't, this tells Alice that her own hat must in fact not be blue. Of course, each player can perform an analogous line of reasoning, so they can all deduce that their own hat is red in time to answer the third question.

[4] Depending on the colour of her own hat, she reasons, Bob and Carl are seeing *at most* one blue hat.

[5] Let's be really careful here to note that, as a matter of fact, Bob does *not* consider it possible that only Carl is wearing a red hat, since he can see a red hat on Alice, too. But Alice doesn't know that! So what we're pointing out is that Alice thinks it's possible *that Bob thinks it's possible* that only Carl has a red hat.

Whew. If you're like me, you might find it a bit difficult at first to follow this argument. Part of the reason is that we're considering *nested* suppositions—specifically, what Alice imagines Bob might be imagining about what Carl sees. English is not terribly good at expressing these types of thoughts clearly.[6]

We can make the reasoning above a lot clearer with the aid of a good diagram. Using a formal structure (like a diagram) to aid or amplify our ability to reason—to offload and encode some important fragment of what we know in a way that enhances our ability to think about it—is, without exaggeration, the essence of mathematics. So let's not be shy![7]

Consider Figure 3.1. It depicts a type of mathematical object called a *graph*, which consists of *nodes*[8] and *edges* connecting pairs of nodes, as shown. Here, additionally, the nodes and edges are labelled.

[6] We managed, but let's be grateful that it wasn't a four or five person game!

[7] As a bonus, the diagram we draw will also be our first introduction to *relational structures*, which are the most common type of models used in modal logic.

[8] Drawn as ellipses.

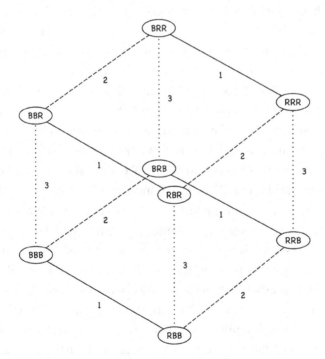

Figure 3.1 The initial state of uncertainty.

Exercise 62

Before reading further, take a minute to study the diagram and make some educated guesses about what the nodes and edges (and their respective labels) represent.

Here's how to think about this picture. Intuitively, the nodes represent all the logically possible configurations of hats worn by the three players, in the order Alice–Bob–Carl. So, for example, the top node, labelled "BRR", represents a situation where Alice has a blue hat and the other two have red hats; meanwhile, the bottom left node, labelled "BBB", represents all players having blue hats.

Next we need to interpret the edges. If you skipped Exercise 62, go back now and do it![9]

The correct answer is "uncertainty" or "indistinguishability" on the part of the player corresponding to the edge's label.[10] So the edges labelled 1 (for Alice) represent Alice's uncertainty via her inability to distinguish between situations where the only difference is the colour of her own hat. For instance, the two nodes mentioned above, "BRR" and "BBB", are *not* connected by an Alice-edge because Alice can tell them apart—in the first, she is seeing a red hat on Bob, while in the second, a blue.[11] By contrast, the distinction between the "BRR" node and the "RRR" node is not one that is apparent to Alice, since she can't see her own hat. Hence the Alice-edge connecting them.

In light of this interpretation of nodes and edges, it's natural to read the edges as representing "epistemic possibility" for the corresponding agent. So, for instance, the Alice-edge considered above connecting "RRR" to "BRR" can be understood as encoding the fact that, in the situation where all three hats are red, Alice considers it possible that her hat is blue (and the other two are red).[12] Similarly, the Bob-edge connecting the "BRR" node to the "BBR" node represents the fact that, in the situation where only Alice is wearing a blue hat, Bob considers it possible

[9] Just stare at the picture for a while with the game in mind, and try to figure out what the edges (and their labels!) represent.

[10] Where of course Alice, Bob, and Carl correspond to the labels 1, 2, and 3, respectively.

[11] This is not the only way she can tell them apart: she is also seeing a red hat on Carl in the first and a blue one in the second.

[12] Since the edges have no inherent "direction", we can equally read this edge in the reverse direction: in the situation where only Alice's hat is blue, she considers it possible that all the hats are red.

that only Carl is wearing a red hat.[13] Moreover, by reading these two edges in succession, we can say that in the situation where everyone is wearing a red hat, Alice considers it possible that Bob considers it possible that only Carl's hat is red.[14]

With this basic structure on the table, we already have a good segue into the general foundations of models for modal logic. Before we move on, however, we should see how we can actually use this diagram to solve the original puzzle![15]

The first step is to figure out how to represent the announcement, "At least one of you is wearing a red hat." Notice that this announcement is incompatible with exactly one node, namely "BBB". Since the announcement is made openly to the whole group, it seems plausible that the effect of the announcement is simply to delete this node from further consideration; see Figure 3.2.

[13] And vice versa.

[14] Compare this to the solution to the puzzle presented above.

[15] This will also serve as a teaser for Chapter 4 on multi-agent logic.

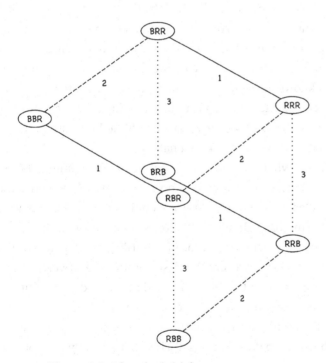

Figure 3.2 After the initial announcement.

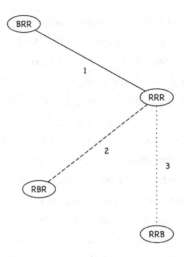

Figure 3.3 After the first question has been answered.

Observe that after this deletion, the "BBR" node now has no Carl-edge connected to it. Since the Carl-edges represent Carl's uncertainty, this corresponds to the fact that in the situation where only Carl's hat is red, the initial announcement *removes* his uncertainty about the colour of his own hat! In other words, Carl knows the colour of his own hat, in this context. Similarly, we see that the "BRB" node is now a situation where Bob knows the colour of his own hat, and the "RBB" node is one where Alice knows the colour of her own hat.

Thus, when the question is asked the first time and everyone answers negatively, it is precisely these three nodes that are ruled out—in each, *someone* would have said "Yes, my hat is red." Since they didn't, those nodes should be eliminated.[16]

Now we see, looking at the updated diagram (Figure 3.3), that after everyone answers the first question negatively, Alice is still unsure about whether all the hats are red, or all but hers are red. However, this non-actual scenario she is entertaining, in which only her hat is blue, has now become one in which Bob knows that his hat is red! This aligns exactly with how we reasoned out the puzzle above. It means that when the question is asked a second time, Bob answering negatively actually

---

[16] One important upshot here is that professions of ignorance can carry useful information! This may help a bit to clarify why asking the same question three times in a row can eventually generate a new answer—the negative answers to the first two questions actually convey new information to the players.

elimates the "BRR" node.[17] In fact, after everyone answers the second question negatively, only one node survives, at which point the diagram shows us clearly that there is *no* uncertainty left, for anyone. They all know their own hat is red.

[17] It also eliminates the "RRB" node. Do you see why?

Exercise 63

(a) Suppose we omit the initial announcement telling the group that at least one of them is wearing a red hat. Use the diagram to argue that in that case, no one will be able to figure out their own hat colour no matter how many times they're asked.

(b) Each person already knows, before it's announced, that there's at least one red hat. How can an announcement of something everyone already knows possibly make a difference in what the players can deduce?

Hint: page 218

What does this example have to do with modal logic? First, the kind of reasoning we appealed to in figuring out the answer to the original puzzle[18] requires a richer language than what classical propositional logic offers—we needed to reason about the knowledge/beliefs of different agents. Second, the diagram we used to aid us in this style of reasoning is a concrete example of what is called *possible world semantics*, utilizing a class of mathematical structures to interpret formulas in modal languages.[19]

[18] For example, "Alice considers it possible that Bob considers it possible that Carl is seeing only one red hat", etc.

[19] Much as valuations are used to interpret formulas in classical propositional logic.

Before diving into formal definitions of the language and the semantics, it's worthwhile to take a moment to emphasize a core mechanic of this "possible worlds" style of representation that is already on display.

We think of each node in the graph—corresponding in this special case to each configuration of hats—as a "world", or in other words, a *way things might be*. Of course, the "way things are" in the real world includes lots of stuff besides hat configurations, but since in this example that's all we're interested in,[20] we restrict our representation to those aspects that concern us.

[20] We don't care whether it's raining, or how tall Alice is, etc.

The edges meanwhile, as discussed, represent uncertainty by representing *possibility*. An Alice-edge connecting world w to world w' means that in world w, Alice considers it possible that the true world might be w'. Because edges represent possibilities, we saw above that a *lack* of edges represents certainty, or knowledge. But we can actually be more general than this: we can say that Alice *knows* φ whenever φ is something that's true in *every world she considers possible*.[21] For instance, looking at the original diagram (Figure 3.1), before anything has been announced, we can say that at the world labelled by "RRR" Alice *knows* that Bob's hat is red. Intuitively, she knows it because she can see it, but the diagram formally represents this in the fact that every world that is connected to "RRR" via an Alice-edge[22] is a world in which Bob is wearing a red hat.

So knowledge, here, can be viewed as corresponding a type of (bounded) universal quantification—namely, Alice knows φ if all the worlds connected to w by an Alice-edge satisfy φ. This turns out to be the core mechanic of standard models for modal logic.

3.2 Language

We extend the classical propositional language in as simple and minimal a way as possible so as to make most salient the particular "modal" phenomena we aim to investigate.[23] The **basic modal language** is specified recursively as follows:[24]

$$\varphi ::= p \mid \neg\varphi \mid \varphi \wedge \psi \mid \Box\varphi,$$

where $p \in$ PROP, a countable[25] set of *primitive propositions*. Let $\mathcal{L}_\Box$ denote the set of all formulas so defined. Other familiar logical connectives like $\vee$ and $\rightarrow$ are defined as abbreviations in the usual way.[26] We also write $\Diamond\varphi$ as an abbreviation for $\neg\Box\neg\varphi$.[27] We refer to modal symbols like "$\Box$", "$\Diamond$", and others we introduce later as *modalities*.

---

[21] Including the actual world.

[22] Including, again, the "RRR" world itself.

[23] We can always expand the language later.

[24] See Chapter 1 for a review of recursive definitions in logic.

[25] See Appendix A.4 for a review of what it means for a set of be "countable".

[26] Specifically, $\varphi \vee \psi$ is an abbreviation for $\neg(\neg\varphi \wedge \neg\psi)$, $\varphi \rightarrow \psi$ is an abbreviation for $\neg(\varphi \wedge \neg\psi)$, and $\varphi \leftrightarrow \psi$ is an abbreviation for $(\varphi \rightarrow \psi) \wedge (\psi \rightarrow \varphi)$.

[27] This is called the *dual* of $\Box$.

Like negation, by convention the box modality always takes *narrow scope* over binary connectives; this means that an expression like $\Box\varphi \wedge \psi$ denotes the formula $(\Box\varphi) \wedge \psi$ and *not* the formula $\Box(\varphi \wedge \psi)$.

The symbol $\Box$ is chosen to look "neutral" since we intend for $\Box\varphi$ to be open to a variety of interpretations.[28] What unites these seemingly disparate interpretations is that they can all be analyzed, mathematically, as a type of *bounded universal quantification*.[29] For instance, "Alice believes φ" can be interpreted as saying something like "φ is true in *all* configurations of the world that Alice considers possible". For another example, "φ will be true henceforth" might naturally be interpreted as saying that "φ is true at *every* point of time in the future". Note that in both cases, the modality is analyzed using words like "all" or "every".

The duality between existential and universal quantification[30] is reflected in the duality between $\Diamond$ and $\Box$. At an intuitive level, dualizing "Alice believes φ" to "Alice does not believe that φ is false" produces a thought that is perhaps more naturally expressed as "Alice considers it possible that φ is true"—corresponding to a bounded *existential* quantifier, namely, "φ is true in *at least one* configuration of the world that Alice considers possible". Similarly, dualizing "φ will be true henceforth" to "φ won't be false henceforth" produces a statement that is arguably better expressed by "φ will be true at *some* future point"—again, a bounded existential quantifier, this time over future points of time.

These ideas are formalized in the definitions of *frames* and *models* and the associated *semantics* for formulas in the basic modal language, which is the topic of the next section. Before moving on, however, it's worth taking a first look at the potential expressive power of modal languages. Even with the relatively simple language defined above,[31] it turns out we have enough expressive power to capture some fairly sophisticated ideas, depending on the interpretation we adopt.

---

[28] For example, "φ is known", "φ is believed", "φ will be true henceforth", "φ ought to be the case", "φ is a guaranteed result (of taking some action)", "φ is knowable", etc.

[29] That is, universal quantification over some restricted set.

[30] That is, the equivalence between $\exists$ and $\neg\forall\neg$.

[31] Extending the classical propositional language with one measly modality.

For instance, suppose we interpret $\Box\varphi$ as "Alice knows that φ", writing it instead as $K\varphi$ to make this intended interpretation more salient. In this case, not only can we clarify our analysis of specific scenarios like the three hats puzzle, but we can also express a variety of properties of (or conjectures about) knowledge more generally.

Consider, for example, the formula $K\varphi \rightarrow \varphi$.[32] This captures what is sometimes called the *factivity* of knowledge, namely, that you can't know false things. More controversial is a formula like $K\varphi \rightarrow KK\varphi$.[33] This is sometimes called *positive introspection*, and represents a certain kind of transparency that knowledge may enjoy or may lack—it's debatable! Syntactically speaking, this property differs from the related notion of *negative introspection*, rendered as $\neg K\varphi \rightarrow K\neg K\varphi$.[34] Does negative introspection follow from positive introspection? What about vice versa? Under what circumstances are these reasonable assumptions about knowledge? These are all questions that arise and that we can address in the context of a formal logic of knowledge.

Now suppose we interpret $\Box\varphi$ as "henceforth, φ", writing it as $H\varphi$ to emphasize this temporal interpretation. Then, as noted above, the dual becomes "at some future point, φ", which we'll write as $F\varphi$. Now consider the formula $HF\varphi$. What does this express? It says that henceforth, there will always be a point of time in the future when φ is true. Another way of thinking of this is that there will be no "last" time when φ is true—it will continue to be true (on and off, perhaps), indefinitely.

If we expand the basic modal language slightly, enough to include two modalities, we can express even more. For instance, a formula like $F(p \wedge Kp)$ says that at some future point p will be true, and at that point Alice will know it. The slightly more complicated formula $F(p \wedge Kp \wedge H(p \rightarrow Kp))$ says that not only will p eventually be true and Alice know it, but also, from that moment on, Alice will continue to know p whenever it's true.

One last example: combining the knowledge modality with a belief modality B allows us to express simple ideas like $K\varphi \rightarrow B\varphi$,

[32] "If Alice knows φ, then it's true."

[33] "If Alice knows φ, then she knows that she knows φ."

[34] "If Alice doesn't know φ, then she knows that she doesn't know it."

that is, that knowledge implies belief, as well as surprising and controverial claims like $B\varphi \to BK\varphi$, that is, if Alice believes φ then she believes that she knows φ.[35]

The point of all these examples (and there are many, many more) is to illustrate that modal languages have the power to express relatively complicated ideas in a variety of domains, making it exceedingly useful to develop tools to reason about them systematically, and specifically, to develop "logics of knowledge", "logics of time", and so on.

[35] Corresponding, perhaps, to a concept of belief as a *feeling of certainty* that cannot be subjectively distinguished from knowledge.

Exercise 64

Suppose we read $K\varphi$ as "Alice knows φ" and $B\varphi$ as "Alice believes φ". Write out readings of the following formulas.

(a) $K(\varphi \wedge \psi) \leftrightarrow (K\varphi \wedge K\psi)$
(b) $B\varphi \to KB\varphi$
(c) $B(B\varphi \to \varphi)$

Exercise 65

Suppose we read $\bigcirc\varphi$ as "tomorrow, φ will be true" and $\square\varphi$ as "φ is necessary"; suppose also that p stands for the proposition that it is raining. Write formulas corresponding the following.

(a) If it rains tomorrow it will also rain the day after tomorrow.
(b) If tomorrow it will necessarily rain, then it is necessary that tomorrow it will rain.
(c) Either it's raining or it's possible that tomorrow it won't rain.

3.3 Semantics: frames and models

The basic challenge of creating semantics for modal logics is the fact that, unlike classical connectives, modalities cannot be reasonably interpreted using truth tables. A truth table for $\square\varphi$

Table 3.1 Truth tables aren't available for belief or henceforth modalities. Compare these with the truth tables for classical connectives presented in Table 1.2.

H	φ	B	φ
?	T	?	T
F?	F	?	F

[36] Depending on whether you interpret "henceforth" to including the current moment or not, one might at least argue that if φ is presently false then it cannot be true henceforth.

[37] See Section 1.2.

[38] Elements of which are called "worlds", or "states", or "points", or "nodes", etc.

[39] For an overview and review of sets and relations, see Appendix A.1.

[40] Note that in contrast to the "hat puzzle" diagram, here the edges are *directed*—in particular, it may be that y is accessible from x but not vice versa.

[41] Recall that 2^X denotes the powerset of X, that is, the set of all subsets of X.

[42] We present the definition in two "pieces" like this—first frames, then models—because we will later wish to consider models without valuations, so it's useful for have a name for that.

[43] Recall that the nodes of the "hat puzzle" diagram were labelled to represent who was wearing what hat. The valuation function plays the same role—it can be viewed as "labelling" each world x with the primitive propositions p that are true at x.

would have to tell us what the truth value of $\Box\varphi$ is *as a function of* the truth value of φ. But none of the interpretations for $\Box$ that we have considered admit such a specification! Whether φ will be true henceforth, for example, is not determined by whether it is true currently;[36] similarly, whether or not Alice believes φ does not depend solely on whether φ is, in fact, true.

So truth tables, which form the basis of the semantics for classical propositional logic,[37] are not enough here—we need a richer structure to capture the intricacies of such modalities. This is where possible world semantics come into play, as illustrated by the hat puzzle. We makes this precise now.

A **frame** is a pair (X, R) where X is a non-empty set[38] and $R \subseteq X \times X$ is a binary *accessibility relation* on X.[39] We write xRy to indicate that $(x, y) \in R$ and in this case say that *y is accessible from x*.[40] For convenience, we also define $R(x) = \{y \in X : xRy\}$, the set of all worlds accessible from x.

A **model** (over PROP) is a frame $F = (X, R)$ together with a *valuation function* $v : \text{PROP} \rightarrow 2^X$.[41],[42] The idea is that a valuation function specifies all the (contingent) truth values of each primitive proposition $p \in \text{PROP}$ at each world $x \in X$—when $x \in v(p)$ we say that *p is true at x*.[43]

Models can be drawn, as we did in the hats puzzle, as (directed) graphs where nodes represent the possible worlds and arrows between nodes represent the accessibility relation. For instance, Figure 3.4 depicts the model $M = (X, R, v)$ where:

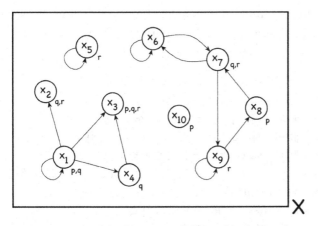

Figure 3.4 A model with 10 possible worlds depicted as a directed graph with nodes representing worlds and edges representing accessibility.

$$X = \{x_1, x_2, \ldots, x_{10}\},$$

$$R = \{(x_1, x_1), (x_1, x_2), (x_1, x_3), (x_1, x_4), (x_4, x_3), (x_5, x_5),$$

$$(x_6, x_6), (x_6, x_7), (x_7, x_6), (x_7, x_9), (x_8, x_7), (x_9, x_9)\},$$

$$v(p) = \{x_1, x_3, x_8, x_{10}\},$$

$$v(q) = \{x_1, x_2, x_3, x_4, x_7\}, \text{ and}$$

$$v(r) = \{x_2, x_3, x_5, x_7, x_9\}.$$

The world-dependent notion of truth for primitive propositions in a model $M = (X, R, v)$ can be extended to all formulas in the basic modal language $\mathcal{L}_\square$ via the following recursive semantic clauses:

$$[\![p]\!]_M = v(p)$$

$$[\![\neg\varphi]\!]_M = X \setminus [\![\varphi]\!]_M$$

$$[\![\varphi \wedge \psi]\!]_M = [\![\varphi]\!]_M \cap [\![\psi]\!]_M$$

$$[\![\square\varphi]\!]_M = \{x \in X : R(x) \subseteq [\![\varphi]\!]_M\}.$$

The set $[\![\varphi]\!]_M$ is sometimes called the *truth set* or the *extension* of φ in M; intuitively, it is the set of worlds in M where φ is true.

Referring back to the model M in Figure 3.4, we list several sample observations to help illustrate various applications of the recursive definition of truth.

- $[\![\neg p]\!]_M = X \setminus [\![\, p\,]\!]_M = \{x_2, x_4, x_5, x_6, x_7, x_9\}$.
- $[\![p \wedge q]\!]_M = [\![\, p\,]\!]_M \cap [\![q]\!]_M = \{x_1, x_3\}$.
- $[\![\neg p \wedge r]\!]_M = [\![\neg p]\!]_M \cap [\![r]\!]_M = \{x_2, x_5, x_7, x_9\}$.
- $x_1 \in [\![\Box q]\!]_M$ since $R(x_1) = \{x_1, x_2, x_3, x_4\} \subseteq [\![q]\!]_M$.
- $x_5 \in [\![\Box r]\!]_M$ since $R(x_5) = \{x_5\} \subseteq [\![r]\!]_M$.
- $x_1 \notin [\![\Box p]\!]_M$ since, for example, $x_1 R x_2$ but $x_2 \notin [\![\, p\,]\!]_M$. It follows that $x_1 \in [\![\neg \Box p]\!]_M$.
- $x_6 \in [\![\Box \neg p]\!]_M$ since $R(x_6) = \{x_6, x_7\} \subseteq [\![\neg p]\!]_M$.
- $x_7 \notin [\![\Box \neg r]\!]_M$ since, for example, $x_7 R x_9$ but $x_9 \notin [\![\neg r]\!]_M$; indeed, we have $[\![\neg r]\!]_M = \{x_1, x_4, x_6, x_8, x_{10}\}$. It follows that $x_7 \in [\![\neg \Box \neg r]\!]_M$.
- $[\![\Box q]\!]_M = \{x_1, x_2, x_3, x_4, x_8, x_{10}\}$ for the following reasons.
  - We have already seen, above, that $x_1 \in [\![\Box q]\!]_M$.
  - We have $R(x_4) = \{x_3\} \subseteq [\![q]\!]_M$, and similarly $R(x_8) = \{x_7\} \subseteq [\![q]\!]_M$.
  - We have $R(x_2) = R(x_3) = R(x_{10}) = \varnothing$, and the empty set is contained in every set, so vacuously $\varnothing \subseteq [\![q]\!]_M$.[44]
- $x_1 \in [\![\Box \Box q]\!]_M$ since $R(x_1) = \{x_1, x_2, x_3, x_4\} \subseteq [\![\Box q]\!]_M$, by the previous bullet point.

[44] Notice that this reasoning is independent of q—any world x with $R(x) = \varnothing$ is a world at which $\Box \varphi$ is (vacuously) true, no matter what φ is!

Exercise 66

(a) Show that in any model M, $[\![\varphi \vee \psi]\!]_M = [\![\varphi]\!]_M \cup [\![\psi]\!]_M$.
(b) Formulate and prove an analogous statement for $[\![\varphi \rightarrow \psi]\!]_M$.

Exercise 67

Compute the following truth sets for the model M in Figure 3.4.

(a) $[\![p \vee q]\!]_M$
(b) $[\![r \rightarrow q]\!]_M$
(c) $[\![r \vee \neg r]\!]_M$

> (d) $[\![\Box q \to \Box p]\!]$
>
> (e) $[\![\Box r \to r]\!]$

The notation $(M, x) \vDash \varphi$ is often used to indicate that $x \in [\![\varphi]\!]_M$, and in this case we say that φ is *true at x* or that *x satisfies φ*.[45] In this terminology, then, we see that $\Box\varphi$ is true at x if and only if $R(x) \subseteq [\![\varphi]\!]_M$, which in turn holds true if and only if φ is true at every world accessible from x. In fact, with these definitions in place, the semantic definitions above can be reformulated as follows:

$$(M, x) \vDash p \text{ iff } x \in v(p)$$

$$(M, x) \vDash \neg\varphi \text{ iff } (M, x) \nvDash \varphi$$

$$(M, x) \vDash \varphi \wedge \psi \text{ iff } (M, x) \vDash \varphi \text{ and } (M, x) \vDash \psi$$

$$(M, x) \vDash \Box\varphi \text{ iff for all } y \in R(x),\ (M, y) \vDash \varphi.$$

Indeed, these equivalences can be taken as an alternative, equivalent definition of truth for formulas in the basic modal language. In this case we can recover the original simply by defining

$$[\![\varphi]\!]_M = \{x \in X \ : \ (M, x) \vDash \varphi\}.$$

Notice that as far as the Boolean connectives are concerned, each world $x \in X$ behaves like a classical model.[46]

Exercise 68

Show that for any model M and any world x, $(M, x) \vDash \varphi \to \psi$ iff $(M, x) \vDash \varphi$ implies $(M, x) \vDash \psi$ (with "implies" here being read as a material conditional).

Next we confirm that the diamond modality really does correspond to bounded existential quantification.[47]

Proposition 31 *Given any world x in any model $M = (X, R, v)$, we have $(M, x) \vDash \Diamond\varphi$ iff there exists a y accessible from x such that $(M, y) \vDash \varphi$. Equivalently,*

$$[\![\Diamond\varphi]\!]_M = \{x \in X \ : \ R(x) \cap [\![\varphi]\!]_M \neq \varnothing\}.$$

[45] When the model is clear from context we sometimes omit it from the notation, writing for example $[\![\varphi]\!]$ or $x \vDash \varphi$. Pairs of the form (M, x) are sometimes called *pointed models*.

[46] In fact, classical propositional logic can be viewed as the special case where there is only one world, $X = \{x\}$. In this case we think of $v(p) = \{x\}$ as p being "true", and $v(p) = \varnothing$ as p being "false"; we ignore the relation because it's only used to interpret the modality, so is not relevant for classical logic.

[47] We are not free to simply *stipulate* the semantic interpretation of $\Diamond\varphi$, since we already defined this formula to be an abbreviation for $\neg\Box\neg\varphi$; this compels us to use the recursive definitions for $\neg$ and $\Box$ given above.

Proof By definition we know that $(M, x) \vDash \Diamond \varphi$ iff $(M, x) \vDash \neg \Box \neg \varphi$. Now the recursive clause for negation implies that this is true iff $(M, x) \nvDash \Box \neg \varphi$, which in turn is equivalent to $R(x) \nsubseteq [\![\neg \varphi]\!]_M$ (using the recursive clause for the box modality). In general, to say that a set A is *not* contained in a set B is equivalent to saying that some point of A is not in B, that is, $A \cap (X \setminus B) \neq \varnothing$.[48] Thus the condition $R(x) \nsubseteq [\![\neg \varphi]\!]_M$ is equivalent to $R(x) \cap (X \setminus [\![\neg \varphi]\!]_M) \neq \varnothing$. And since by definition $[\![\neg \varphi]\!]_M = X \setminus [\![\varphi]\!]_M$, we see that

$$X \setminus [\![\neg \varphi]\!]_M = X \setminus (X \setminus [\![\varphi]\!]_M) = [\![\varphi]\!]_M,$$

so the condition simplifies to $R(x) \cap [\![\varphi]\!]_M \neq \varnothing$, as desired. $\qquad \square$

[48] This is where we are implicitly using the duality of $\exists$ and $\forall$. Do you see it? Hint: set containment is essentially a universal statement.

Exercise 69

Prove the following claims about the model M in Figure 3.4.

(a) $(M, x_2) \vDash q$ but $(M, x_2) \nvDash \Diamond q$.
(b) $(M, x_4) \vDash \Diamond(p \wedge q \wedge r)$ but $(M, x_4) \nvDash \Diamond \Diamond r$.
(c) $(M, x_7) \nvDash \Diamond q$ but $(M, x_7) \vDash \Diamond \Diamond q$.
(d) $(M, x_8) \nvDash \Diamond p$ and $(M, x_8) \nvDash \Diamond \Diamond p$, but $(M, x_8) \vDash \Diamond \Diamond \Diamond p$.
(e) $(M, x_9) \vDash \Diamond(p \wedge \Diamond q)$, but $(M, x_9) \nvDash \Diamond(p \wedge q)$.
(f) $(M, x_1) \vDash \Diamond \Box p$ but $(M, x_1) \nvDash \Box \Diamond p$.

Exercise 70

Give examples of worlds x in models M satisfying each of the following (or prove that no such world–model pair exists).

(a) $\Diamond p \wedge \neg \Box p$
(b) $\Box p \wedge \neg \Diamond p$
(c) $\Diamond p \wedge \Diamond q \wedge \neg \Diamond(p \wedge q)$
(d) $\Box p \wedge \Box q \wedge \neg \Box(p \wedge q)$
(e) $\Box p \wedge \neg p$
(f) $\Diamond \Diamond p \wedge \Box \neg p$
(g) $\Box \Diamond p \wedge \Diamond \Box p$

A formula φ is said to be **valid in a model** M if $[\![\varphi]\!]_M = X$; that is, if φ is true at *every* world in M. In this case we write $M \vDash \varphi$. Conversely, if $M \nvDash \varphi$ then we say that M *refutes* φ.[49]

[49] In this case, there must be some world x in M such that $(M, x) \nvDash \varphi$, and we also say that M *refutes* φ *at* x.

Exercise 71

Suppose $M = (W, R, v)$ is a model with worlds $W = \{a, b, c, d\}$, accessibility relation $R = \{(a,b), (b,c), (c,a), (d,a), (d,b), (d,c)\}$, and valuation v satisfying $v(p) = \{a, c\}$ and $v(q) = \{a, b\}$. Draw a picture of this model and evaluate the following claims.

(a) $(M, a) \vDash \Box p$
(b) $(M, a) \vDash \Diamond\Diamond p$
(c) $(M, d) \vDash \Diamond\Box\Box p$
(d) $M \vDash \neg q \vee p$
(e) $M \vDash \Box(p \vee q)$
(f) $M \vDash \Diamond q \rightarrow \Box q$

Intuitively, the truth of a formula that is valid in M doesn't depend on the particular world. However, it is still tied to the arbitrariness of the valuation in that model. Just as in classical propositional logic, we are interested in identifying which formulas are true in virtue of their form, rather than merely contingently true; that is, we care about formulas that are true regardless of v.

To this end, we say a formula φ is **valid in a frame** F if $M \vDash \varphi$ for every model M based on F.[50] In this case we write $F \vDash \varphi$.[51] So, as in classical propositional logic, we remove the dependence on v by "quantifying v out".

But we can go farther than this. A formula might be valid on a frame F in virtue of specific facts about that frame, and we may similarly wish to abstract away from such specifics. Thus we say a formula φ is **valid in a class of frames** $\mathcal{F}$ if $F \vDash \varphi$ for all $F \in \mathcal{F}$.[52] In this case we write $\mathcal{F} \vDash \varphi$. At the extreme, a formula φ is said to be simply **valid** if it is valid on every frame. In this case we write $\vDash \varphi$.

[50] The model (X, R, v) is said to be *based on* the frame (X, R).

[51] As above, when $F \nvDash \varphi$ we say that F *refutes* φ.

[52] For instance, we may be interested in which formulas are valid on the class of frames where the relation R is reflexive, or where the set of worlds is finite.

Proposition 32 $\vDash \varphi \vee \neg\varphi$.

Proof From the definition of validity, what we must show is that in any model $M = (X, R, v)$, we have

$$[\![\varphi \vee \neg\varphi]\!]_M = X.$$

In light of Exercise 66, this is equivalent to showing that

$$[\![\varphi]\!]_M \cup [\![\neg\varphi]\!]_M = X.$$

And since $[\![\neg\varphi]\!]_M$ is (again, by definition) just the complement of $[\![\varphi]\!]_M$, this is obvious. $\qquad\square$

As explored in Exercise 66, the recursive formula for $[\![\varphi \rightarrow \psi]\!]_M$ does not look terribly natural, owing to the somewhat awkward definition of the material conditional in terms of conjunction and negation. However, at the level of *validity*, material implication has an exceedingly natural formulation.

Lemma 33 $M \vDash \varphi \rightarrow \psi$ *if and only if* $[\![\varphi]\!]_M \subseteq [\![\psi]\!]_M$.

[53] Here and elsewhere, when the model is presented simply with the letter "M", we assume for convenience that the state space, accessibility relation, and valuation are denoted by X, R, and v, respectively.

Proof By definition $M \vDash \varphi \rightarrow \psi$ just in case $[\![\varphi \rightarrow \psi]\!]_M = X$,[53] which holds exactly when

$$X \setminus ([\![\varphi]\!]_M \cap (X \setminus [\![\psi]\!]_M)) = X.$$

This simplifies to

$$[\![\varphi]\!]_M \cap (X \setminus [\![\psi]\!]_M) = \varnothing,$$

which in turn is equivalent to $[\![\varphi]\!]_M \subseteq [\![\psi]\!]_M$. $\qquad\square$

Corollary 34 $M \vDash \varphi \leftrightarrow \psi$ *if and only if* $[\![\varphi]\!]_M = [\![\psi]\!]_M$.

Next we encounter our first example of a valid formula that is *not* a propositional tautology.[54]

[54] Read the proof carefully: nothing deep is happening, but it intentionally mixes the various notations we have developed so far, to show how they can be freely interwoven.

Proposition 35 $\vDash \Box(\varphi \rightarrow \psi) \rightarrow (\Box\varphi \rightarrow \Box\psi)$.

Proof Let $M = (X, R, v)$ be an arbitrary model; we must show that

$$M \vDash \Box(\varphi \rightarrow \psi) \rightarrow (\Box\varphi \rightarrow \Box\psi).$$

By Lemma 33, this is equivalent to showing that

$$[\![\Box(\varphi \rightarrow \psi)]\!] \subseteq [\![(\Box\varphi \rightarrow \Box\psi)]\!].$$

So let $x \in [\![\Box(\varphi \rightarrow \psi)]\!]$, which means $x \vDash \Box(\varphi \rightarrow \psi)$; we wish to show that $x \in [\![\Box\varphi \rightarrow \Box\psi]\!]$, that is, that $x \vDash \Box\varphi \rightarrow \Box\psi$. To show the latter, by Exercise 68 we need to show that if $x \vDash \Box\varphi$ then $x \vDash \Box\psi$.

So suppose that $x \vDash \Box\varphi$. This means that $R(x) \subseteq [\![\varphi]\!]$. Since we are also supposing that $x \vDash \Box(\varphi \rightarrow \psi)$, we also have $R(x) \subseteq [\![\varphi \rightarrow \psi]\!]$. Putting these facts together, we see that for every $y \in R(x)$, we have $y \vDash \varphi$ and $y \vDash \varphi \rightarrow \psi$. Thus, for each such y we also have $y \vDash \psi$, so $R(x) \subseteq [\![\psi]\!]$, whence $x \vDash \Box\psi$ by definition. $\square$

Exercise 72

Show that the following formulas are valid.

(a) $\Box\varphi \rightarrow \Box(\varphi \vee \psi)$
(b) $\Box(\varphi \wedge \psi) \leftrightarrow (\Box\varphi \wedge \Box\psi)$
(c) $\Diamond\varphi \leftrightarrow \neg\Box\neg\varphi$
(d) $\Box\varphi \leftrightarrow \neg\Diamond\neg\varphi$

Solution: page 218

Now let's see an example of a formula that's *not* valid.

Proposition 36 $\nvDash \Box p \rightarrow p$.

Proof Since this is the negation of a universal claim,[55] it suffices to find a counterexample. Specifically, we wish to exhibit a model that doesn't validate $\Box p \rightarrow p$.

Consider the model $M = (X, R, v)$ where $X = \{a, b\}$, $R = \{(a, b)\}$, and $v(p) = \{b\}$.[56] Then we can see that $a \vDash \Box p$, since $R(a) = \{b\} \subseteq [\![p]\!]$, but clearly $a \nvDash p$, so $a \nvDash \Box p \rightarrow p$. $\square$

Comparing Propositions 35 and 36, you may notice that in the former we used *metavariables* φ and ψ (ranging over all formulas in $\mathcal{L}_\Box$), whereas in the latter we restricted attention to a particular primitive proposition p. And indeed, in the former,

[55] Namely, the claim that *all* models validate $\Box p \rightarrow p$.

[56] This can be drawn as usual, as follows.

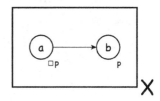

we proved validity for *all* formulas φ and ψ. This is yet another, somewhat hidden, universal quanitifier at play when we reason about validity: the expression

$$\vDash \Box(\varphi \rightarrow \psi) \rightarrow (\Box\varphi \rightarrow \Box\psi)$$

asserts that for *all* formulas $\varphi, \psi \in \mathcal{L}_\Box$, the corresponding formula is valid. Similarly, the expression

$$\vDash \Box\varphi \rightarrow \varphi$$

asserts that for all formulas $\varphi \in \mathcal{L}_\Box$, $\Box\varphi \rightarrow \varphi$ is valid. Thus, finding even a single formula φ for which this fails is enough to refute the claim. So if we simply take φ to be the formula p and appeal to Proposition 36, we immediately obtain:

Corollary 37 $\nvDash \Box\varphi \rightarrow \varphi$.

Exercise 73

Show that the following formulas are *not* valid.

(a) $\Box(\varphi \vee \psi) \rightarrow (\Box\varphi \vee \Box\psi)$
(b) $\Diamond(\varphi \wedge \psi) \leftrightarrow (\Diamond\varphi \wedge \Diamond\psi)$
(c) $\varphi \rightarrow \Box\Diamond\varphi$
(d) $\Diamond\Box\varphi \rightarrow \Box\Diamond\varphi$

Solution: page 218

Looking back at the proof that $\Box p \rightarrow p$ is not valid, we can see that the counterexample worked because there was a world x that could only access p-worlds, yet itself was not a p-world. This, of course, would be impossible if x was accessible from itself; this insight leads us to consider the class $\mathcal{F}_{ref}$ of *reflexive frames*, that is, those frames (X, R) where the relation R is reflexive.[57]

[57] Recall that a relation is called *reflexive* if, for all x, xRx. See Appendix A.1.

Proposition 38 $\mathcal{F}_{ref} \vDash \Box\varphi \rightarrow \varphi$.

Proof Let $M = (X, R, v)$ be an arbitrary model based on a reflexive frame, and let $x \in X$. We wish to show that $x \vDash \Box\varphi \rightarrow \varphi$, or equivalently, that whenever $x \vDash \Box\varphi$, also $x \vDash \varphi$.

So suppose that $x \vDash \Box\varphi$. Then by definition we have $R(x) \subseteq [\![\varphi]\!]$. Moreover, since R is reflexive, we know that $x \in R(x)$. Putting these facts together yields $x \in [\![\varphi]\!]$, so $x \vDash \varphi$, as desired. □

Exercise 74

Call a frame (X, R) *trivial* if $R = \varnothing$. For each formula in Exercise 73, explain whether or not it is valid with respect to the class of trivial frames, and find a (non-empty) class of *non-trivial* frames on which it is valid.

Solution: page 219

It turns out that many natural properties of frames correspond to (relatively) simple modal formulas in this way. For example, it's not too hard to see (Proposition 39) that although the formula $\Box\varphi \to \Box\Box\varphi$ is not valid in general, it *is* valid on the class $\mathcal{F}_{trans}$ of transitive frames.[58]

This sort of connection between logical formulas and graph-theoretic properties like reflexivity or transitivity is one of the core enterprises of modal logic: in the epistemic context, for example, it establishes a correspondence between properties of knowledge[59] and properties of relations,[60] opening the door to studying epistemic questions using tools from graph theory!

Proposition 39 $\nvDash \Box\varphi \to \Box\Box\varphi$, but $\mathcal{F}_{trans} \vDash \Box\varphi \to \Box\Box\varphi$.

Proof First we show the formula is not valid by constructing a counterexample. Take $\varphi = p$; then we want to build a world x that satisfies $\Box p$ but not $\Box\Box p$, so we need to arrange for all the worlds accessible from x to be p-worlds, but for at least one world accessible from x to *not* be a $\Box p$-world.

Perhaps the most straightforward way to arrange this is with the model (X, R, v) where $X = \{a, b, c\}$, $R = \{(a, b), (b, c)\}$, and $v(p) = \{b\}$.[61] Then we can see that $a \vDash \Box p$ (since $R(a) = \{b\} \subseteq [\![p]\!]$), but $b \nvDash \Box p$, since $R(b) \nsubseteq [\![p]\!]$. Thus $a \nvDash \Box p \to \Box\Box p$, as desired.

Clearly this model is not transitive—if it were, we would also have aRc, in which case we would no longer have $a \vDash \Box p$. And if

[58] Recall that a relation R is transitive if for all x, y, and z, whenever xRy and yRz, also xRz. See Appendix A.1.

[59] Like "you only know true things" ($K\varphi \to \varphi$) or "anything you know, you also know that you know" ($K\varphi \to KK\varphi$).

[60] Like reflexivity or transitivity.

[61] Pictured here.

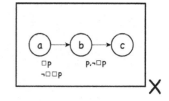

we tried to "fix" this by redefining $v(p) = \{b,c\}$, then we would no longer have $b \nvDash \Box p$.[62] This is no accident; in fact, we will now show that *any* transitive model M validates $\Box\varphi \to \Box\Box\varphi$.

Let M be a transitive model and let x be a point in that model. Suppose $x \vDash \Box\varphi$. We wish to show that $x \vDash \Box\Box\varphi$, or in other worlds, that for every $y \in R(x), y \vDash \Box\varphi$. To show this for any given $y \in R(x)$, we must show that all $z \in R(y)$ are such that $z \vDash \varphi$. But observe that in this case we have xRy and yRz, so by transitivity, also xRz. Therefore, since $x \vDash \Box\varphi$ by assumption, we must have $z \vDash \varphi$, as desired. □

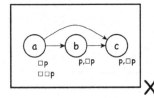

[62] Pictured here.

✗

Exercise 75

Prove the following.

(a) $\mathcal{F}_{ref} \vDash \varphi \to \Diamond\varphi$
(b) $\mathcal{F}_{ref} \nvDash \Box\varphi \to \Box\Box\varphi$
(c) $\mathcal{F}_{trans} \vDash \Diamond\Diamond\varphi \to \Diamond\varphi$
(d) $\mathcal{F}_{trans} \nvDash \neg\Box\varphi \to \Box\neg\Box\varphi$

Exercise 76

Consider the formula $\Box\varphi \to \Box\Diamond\varphi$. For each of the following, justify your answer.

(a) Is this formula valid on the class of all reflexive frames?
(b) Is this formula valid on the class of all symmetric frames?

At this point one might notice a certain pattern: not only does $\mathcal{F}_{ref}$ validate $\Box\varphi \to \varphi$, but also $\varphi \to \Diamond\varphi$; not only does $\mathcal{F}_{trans}$ validate $\Box\varphi \to \Box\Box\varphi$, but also $\Diamond\Diamond\varphi \to \Diamond\varphi$. This is no coincidence: the hidden universal quantifier in the claim that $\mathcal{F}_{ref} \vDash \Box\varphi \to \varphi$, for instance, means that we also immediately have $\mathcal{F}_{ref} \vDash \Box\neg\varphi \to \neg\varphi$; but this formula is equivalent to its contrapositive, so $\mathcal{F}_{ref} \vDash \neg\neg\varphi \to \neg\Box\neg\varphi$, which this in turn yields $\mathcal{F}_{ref} \vDash \varphi \to \Diamond\varphi$. Similar reasoning gives us:

$$\mathcal{F}_{trans} \vDash \Box\varphi \to \Box\Box\varphi \text{ iff } \mathcal{F}_{trans} \vDash \Box\neg\varphi \to \Box\Box\neg\varphi$$

$$\text{iff } \mathcal{F}_{trans} \vDash \neg\Box\Box\neg\varphi \to \neg\Box\neg\varphi$$

$$\text{iff } \mathcal{F}_{trans} \vDash \neg\Box\neg\neg\Box\neg\varphi \to \Diamond\varphi$$

$$\text{iff } \mathcal{F}_{trans} \vDash \Diamond\Diamond\varphi \to \Diamond\varphi.$$

So, roughly speaking, when it comes to validity in this context one can always "flip the implication and switch boxes and diamonds" to get an equivalent valid formula. Sometimes this can really help with intuition and proofs!

The more we restrict a class of frames, the more formulas it might validate.[63] Given a formula φ, we might then try to find the *least restrictive* class of frames $\mathcal{F}$ that validates φ. This idea is formalized in the following definition.

A formula φ **defines** a class of frames $\mathcal{F}$ if, for every frame F,

$$F \vDash \varphi \text{ iff } F \in \mathcal{F}.$$

So the right-to-left direction is simply the requirement that $\mathcal{F} \vDash \varphi$, whereas the left-to-right requirement demands (in contrapositive) that whenever $F \notin \mathcal{F}$, some model based on F refutes φ.[64]

For example, although $\mathcal{F}_{ref} \vDash p \to p$, this formula does not *define* $\mathcal{F}_{ref}$ since it is also valid on frames outside of this class.[65] By contrast, the formula $\Box p \to p$ is not just valid on $\mathcal{F}_{ref}$ but in fact defines this class:

Proposition 40 *The formula $\Box p \to p$ defines the class of reflexive frames.*

Proof We have already established (Proposition 38) that $\mathcal{F}_{ref} \vDash \Box p \to p$. So what remains to show is that if $F \notin \mathcal{F}_{ref}$, then $F \nvDash \Box p \to p$. Essentially we must show that any non-reflexive frame F can be equipped with a valuation function in such a way that the resulting model refutes $\Box p \to p$.[66]

So let $F = (X, R)$ be a non-reflexive frame. That means it contains at least one point x such that $x \notin R(x)$. This is the point at which we will refute $\Box p \to p$; to do so, we simply need to define

[63] One way of thinking about this is that a more restricted class of frames has fewer counterexamples available.

[64] To prove that *some* model based on F refutes φ, we just need to find a valuation that, when attached to F, produces a model that refutes φ.

[65] In fact, this formula is valid on *all* frames, so (trivially) it defines the class of all frames.

[66] Remember that to prove $F \nvDash \varphi$, we need to show that *some* model based on F refutes φ. That's why we get to choose the valuation function—as long as we can find any model based on F that refutes φ, we're done.

a valuation v that makes $\Box p$ true at x in the resulting model, and p false at x. One way to do this is simply to define $v(p) = X\backslash\{x\}$.[67] This certainly ensures that $x \nvDash p$; moreover, since $x \notin R(x)$ we know that $R(x) \subseteq X \backslash \{x\}$, from which it follows that $x \vDash \Box p$, as desired. $\Box$

So we have shown that the correspondence between the formula $\Box p \to p$ and the frame property of reflexivity is "tight", in the sense that reflexivity is strong enough to ensure the validity of this formula, and no weaker property would suffice to do so.[68]

It's helpful to consider a contrasting example: $\Box p \to p$ does *not* define the class $\mathcal{F}_{id}$.[69] Although it's easy to see that $\mathcal{F}_{id} \vDash \Box p \to p$, nonetheless this formula does not define $\mathcal{F}_{id}$ because one can find frames outside this class that also validate $\Box p \to p$.

Exercise 77

Prove that $\Box p \to p$ does not define the class $\mathcal{F}_{id}$.

Let's look at one final example of definability in detail.

Proposition 41 $\Box p \to \Box\Box p$ *defines* $\mathcal{F}_{trans}$.

Proof We have already seen (Proposition 39) that $\mathcal{F}_{trans} \vDash \Box p \to \Box\Box p$. For the converse, we must show that any non-transitive frame F can be equipped with a valuation function in such a way that the resulting model refutes $\Box p \to \Box\Box p$.

So let $F = (X, R)$ be a non-transitive frame. This means we can find points $x, y, z \in X$ such that xRy and yRz but $z \notin R(x)$.[70] We'll refute the formula at x; this means we need a valuation that guarantees that $x \vDash \Box p$ but $x \nvDash \Box\Box p$. To arrange the latter, since z is the only point we know to be accessible from a point accessible from x, we better make sure that $z \nvDash p$. Let us therefore define $v(p) = X \backslash \{z\}$.[71]

We can see that in the resulting model, since yRz and $z \nvDash p$, we have $y \nvDash \Box p$. Thus, since xRy, we obtain $x \nvDash \Box\Box p$, which is half of what we wanted. It remains to show that $x \vDash \Box p$, or in

---

[67] We could also have defined $v(p) = R(x)$. Do you see why?

[68] It is crucial to note here that we are speaking of validity *in frames*, not in models. For example, it is relatively easy to come up with a non-reflexive model in which $\Box p \to p$ is valid (say, when $v(p) = X$). The valuation can often be "rigged", so to speak, to make a formula valid even if it could easily be refuted by a different valuation. Frame definability, by contrast, is about capturing properties of frames that hold independent of the valuation.

[69] That is, the class of all frames where R is the identity relation, so xRy iff $x = y$.

[70] Note well: there is no guarantee that these points are all distinct! The conditions in this case imply that $x \neq y$ and $y \neq z$ (why?), but they are compatible with $x = z$!

[71] We could have also defined $v(p) = R(x)$. Do you see why?

other words that $R(x) \subseteq v(p)$. But this follows immediately from the fact that $z \notin R(x)$, since z is the only world not in $v(p)$. $\square$

Exercise 78

Let

$$\mathcal{F}_{par} = \{(X, R) : \forall x(|R(x)| \leq 1)\},$$

that is, the class of frames where there is at most one world accessible from any given world. Show that $(\Diamond p \wedge \Diamond q) \rightarrow \Diamond(p \wedge q)$ defines $\mathcal{F}_{par}$.

Exercise 79

Prove that a formula can define at most one class of frames.

Exercise 80

Find the class of frames defined by each of the following formulas. Justify your answers.

(a) $\Box p \rightarrow \Diamond p$
(b) $\Box(\Box p \rightarrow p)$
(c) $\Diamond\Box p \rightarrow \Box\Diamond p$

Solution: page 219

Exercise 81

For each of the following classes of frames, find a formula that defines it.

(a) The class of frames where R is the identity relation, $\mathcal{F}_{id}$.
(b) The class of *symmetric* frames:

$$\mathcal{F}_{sym} = \{(X, R) : \forall x, y(xRy \text{ implies } yRx)\}.$$

(c) The class of *Euclidean* frames:

$$\mathcal{F}_{eucl} = \{(X, R) : \forall x, y, z((xRy \text{ and } xRz) \text{ implies } yRz)\}.$$

Solution: page 220

Exercise 82

Consider a modal language with two (box-type) modalities K and B (read as "the agent knows" and "the agent believes", respectively) interpreted in relational structures with two relations R_K and R_B in the usual way.

(a) Consider the formula $Kp \to Bp$ ("If the agent knows p then she believes p.") Find the class of frames defined by this formula. Justify your answer.
(b) Consider the formula $Bp \to BKp$ ("If the agent believes p, then she believes that she knows p."). Find the class of frames defined by this formula. Justify your answer.

Solution: page 221

There are many classes of frames that are *not* defined by any formula in the basic modal language. To prove such *non-definability* results, we make use of a concept called "invariance".[72]

3.4 Invariance and expressivity

Let x be a world in some model M. The **(basic modal) theory of x in M** is the set

$$Th_M(x) = \{\varphi \in \mathcal{L}_\Box \; : \; (M, x) \vDash \varphi\}.$$

In other words, it's the set of all formulas in the basic modal language that are true at x. There can be different worlds that have the *same theory*, and the idea of invariance is to investigate ways of recognizing when this happens.

Perhaps the simplest type of invariance is *invariance under disjoint union*. Given two models $M_1 = (X_1, R_1, v_1)$ and $M_2 = (X_2, R_2, v_2)$ with $X_1 \cap X_2 = \varnothing$,[73] define the **disjoint union of M_1 and M_2** to be the model $M_1 \sqcup M_2 = (X, R, v)$ where:[74]

- $X = X_1 \cup X_2$;
- $R = R_1 \cup R_2$;
- $v(p) = v_1(p) \cup v_2(p)$.

[72] The following section is more mathematically dense than what we've seen so far in this chapter. It can be safely skimmed or skipped on a first read-through.

[73] If you happen to start with two models where $X_1 \cap X_2 \neq \varnothing$, there is a common trick to "disjointify" them artificially: replace each $x \in X_1$ with the pair $(x, 1)$, and each $y \in X_2$ with $(y, 2)$. Intuitively, each point now "remembers" which set it originally came from.

[74] We can also define a *disjoint union of frames* by simply ignoring the condition on the valuation.

Roughly speaking, $M_1 \sqcup M_2$ is what you get when you combine the separate models M_1 and M_2 together into a single model *without messing up* any of the structure these models had independently. More precisely, what this means is cashed out exactly by the definition above: all the worlds of each model are preserved (and remain distinct); all the relations between points are likewise preserved,[75] and the valuation assigns to each world in $M_1 \sqcup M_2$ the same primitive propositions that were true at that world in its original model.

So, intuitively, although we have created a new, bigger model out of two old models, each world within the new model still "thinks" it is in the old, smaller model it came from. This is formalized in the following.

Proposition 42 *For all $x \in X_1$, $Th_{M_1}(x) = Th_{M_1 \sqcup M_2}(x)$; likewise, for all $y \in X_2$, $Th_{M_2}(y) = Th_{M_1 \sqcup M_2}(y)$.*

Proof We prove the first result (the second is completely analogous). What we wish to show can be equivalently formulated as follows: *for all $\varphi \in \mathcal{L}_\square$, for all $x \in X_1$, $(M_1, x) \vDash \varphi$ iff $(M_1 \sqcup M_2, x) \vDash \varphi$.* We can prove this by structural induction on φ.[76]

For the base case when $\varphi = p \in$ PROP, we must show that for all $x \in X_1$, $(M_1, x) \vDash p$ iff $(M_1 \sqcup M_2, x) \vDash p$. This is equivalent to showing that $x \in v_1(p)$ iff $x \in v(p)$, which follows directly from the definition of v.[77]

Now suppose inductively that the result holds for φ and ψ; we wish to show it holds for $\neg\varphi$, $\varphi \wedge \psi$, and $\square\varphi$. The first two cases are trivial, since the truth conditions at x for Boolean connectives don't depend at all on the structure of the rest of the model.[78] So we focus on the final inductive step, corresponding to the box modality: we wish to show that for all $x \in X_1$,

$$(M_1, x) \vDash \square\varphi \text{ iff } (M_1 \sqcup M_2, x) \vDash \square\varphi.$$

By definition, this is equivalent to showing that for each $x \in X_1$,

$$R_1(x) \subseteq \llbracket \varphi \rrbracket_{M_1} \text{ iff } R(x) \subseteq \llbracket \varphi \rrbracket_M. \tag{3.1}$$

[75] And no new relations are added, so in particular this means there are never any edges in $M_1 \sqcup M_2$ connecting (former) points of M_1 with those of M_2.

[76] See Chapter 1 for a review of structural induction.

[77] And the fact that X_1 and X_2 are disjoint.

[78] Write out these inductive steps carefully to convince yourself of this.

From the definition of R we know that for any $y \in X$, xR_1y iff xRy, and thus $R_1(x) = R(x) = R(x) \cap X_1$. Moreover, the inductive hypothesis tells us that for all $y \in X_1$, $(M_1, y) \vDash \varphi$ iff $(M_1 \sqcup M_2, y) \vDash \varphi$, and therefore $[\![\varphi]\!]_{M_1} = [\![\varphi]\!]_M \cap X_1$.[79] Thus we have

$$R_1(x) \subseteq [\![\varphi]\!]_{M_1} \text{ iff } R(x) \cap X_1 \subseteq [\![\varphi]\!]_M \cap X_1$$

$$\text{iff } R(x) \subseteq [\![\varphi]\!]_M,$$

which establishes the equivalence in (3.1) and completes the induction.   □

It is perhaps not terribly surprising that simply "pushing two models beside each other" doesn't change the truth values of any formulas in those models. A more refined notion of invariance—one that takes specific advantage of the way modality is interpreted via an accessibility relation—is called *invariance of generated submodels*.

The basic idea here is that, in evaluating any given modal formula φ in a model M, we only ever need to look "forward" along the relation R. For instance, if φ is $\Box p$ or $\Diamond p$, then to determine the truth of φ at x we must look at worlds in $R(x)$; similarly, if φ is $\Box\Box p$ or, say, $\Diamond\Box p$, we must look at worlds accessible from worlds in $R(x)$. But no matter how deeply the modalities in φ are nested, we will never be forced to look beyond those worlds that can be reached via a succession of R-edges. This concept is at the core of defining a generated submodel.

We first need to capture this notion of worlds that can be reached "via a succession of R-edges". Given a relation R on a set X, an *R-path* is a finite sequence of points $x_1, \ldots, x_k$ such that for all $1 \leq i < n$, x_iRx_{i+1}. Say that y is *R-reachable from* x if there exists an R-path $x_1, \ldots, x_k$ with $x_1 = x$ and $x_k = y$. Define

$$R^* = \{(x, y) : y \text{ is } R\text{-reachable from } x\}.$$

This relation R^* is called the *reflexive–transitive closure* of R because it happens to be the smallest extension of R that is both reflexive and transitive.

[79] There may be φ-worlds in M that aren't in X_1, of course, which is why we need this intersection.

Proposition 43 R^* *is the smallest reflexive and transitive relation containing R.*

Proof First, $R \subseteq R^*$ because if $(x, y) \in R$, then y is R-reachable from x via the (two-step) R-path x, y, so by definition xR^*y.

Next we show that R^* is reflexive and transitive. Reflexivity holds because every x is R-reachable from itself via the (one-step) R-path x. For transitivity, observe that if y is R-reachable from x via the R-path $x_1, \ldots, x_k$ and z is R-reachable from y via the R-path $y_1, \ldots, y_\ell$, then z is R-reachable from x via the R-path $x_1, \ldots, x_k, y_2, \ldots, y_\ell$.

Finally, to see that R^* is the *smallest* such relation, let S be an arbitrary reflexive and transitive relation containing R. We'll show that $R^* \subseteq S$. Let $(x, y) \in R^*$; then y is R-reachable from x, and since S contains R, we know that y is also S-reachable from x. Because S is transitive (and reflexive), a simple induction on the length of the S-path from x to y will establish that $(x, y) \in S$, which will complete the proof.

Here is the induction. We prove the following claim: *for all $k \geq 1$, if $x_1, \ldots, x_k$ is an S-path, then $x_1 S x_k$.* In the base case $k = 1$ we have $x_1 = x_k$, so $x_1 S x_k$ follows by reflexivity. Now suppose the result holds for k; we want to show it holds for $k + 1$. So suppose that $x_1, \ldots, x_k, x_{k+1}$ is an S-path. Then clearly $x_1, \ldots, x_k$ is an S-path as well, so the inductive hypothesis applies to tell us that $x_1 S x_k$. Since also $x_k S x_{k+1}$, transitivity of S implies that $x_1 S x_{k+1}$, as desired. $\square$

Exercise 83

Extend the basic modal language with a new modality $\square^*$, interpreted using R^* in the usual way.

(a) Show that $\models \square^* \varphi \to \varphi$ and $\models \square^* \varphi \to \square^* \square^* \varphi$.
(b) Show that $\models \square^*(\varphi \to \square \varphi) \to (\varphi \to \square^* \varphi)$.

Solution: page 221

We will use this notion of reachability to essentially "throw away" every point in a model that is not reachable from some given point x, and show that in the resulting, smaller model, the theory of x remains unchanged.[80]

[80] In fact, we'll show that the theory of every point reachable from x also remains the same.

Given a model $M = (X, R, v)$ and a subset $A \subseteq X$, define the *restriction of R to A*, denoted $R|_A$, by

$$R|_A = \{(x, y) \in A^2 \ : \ xRy\},$$

and likewise the *restriction of v to A*, denoted $v|_A$, by

$$v|_A(p) = v(p) \cap A.$$

So in both cases we are simply ignoring points outside of A.

This brings us at last to the promised definition: given $x \in X$, define the **generated submodel of M at x** to be the model[81]

[81] We can also define a *generated subframe* by simply ignoring the valuation.

$$M_x = (R^*(x), R|_{R^*(x)}, v|_{R^*(x)}).$$

To avoid notational clutter, we will sometimes write $M_x = (X_x, R_x, v_x)$.

Proposition 44 *For all $y \in R^*(x)$, $\mathit{Th}_{M_x}(y) = \mathit{Th}_M(y)$.*

Proof As in Proposition 42, what we wish to prove is equivalent to the following: *for all $\varphi \in \mathcal{L}_\Box$, for all $y \in R^*(x)$, $(M, y) \vDash \varphi$ iff $(M_x, y) \vDash \varphi$.* We proceed by structural induction on φ.

The base case of the induction follows simply from the definition of v_x, since for all $y \in R^*(x)$ and any $p \in \text{PROP}$ it is easy to see that $y \in v(p)$ iff $y \in v_x(p)$, and thus $(M, y) \vDash p$ iff $(M_x, y) \vDash p$.

The inductive steps for $\neg$ and $\wedge$ are trivial.[82] Finally, the inductive step for $\Box$ follows from the fact that, for all $y \in R^*(x)$, $R(y) = R_x(y)$,[83] and $\llbracket \varphi \rrbracket_{M_x} = \llbracket \varphi \rrbracket_M \cap R^*(x)$, which follows from the inductive hypothesis. Indeed, we then have:

[82] As in Proposition 42.

[83] So, intuitively, $\Box$ behaves the same way in M_x as it did in M.

$$(M, y) \vDash \Box\varphi \text{ iff } R(y) \subseteq \llbracket \varphi \rrbracket_M$$
$$\text{iff } R_x(y) \subseteq \llbracket \varphi \rrbracket_M$$
$$\text{iff } R_x(y) \subseteq \llbracket \varphi \rrbracket_{M_x}$$
$$\text{iff } (M_x, y) \vDash \Box\varphi.$$

This completes the induction. $\qquad\qquad\qquad\qquad\qquad\square$

Much more general than either invariance of disjoint unions or invariance of generated submodels is *invariance under bisimulation.* Given two models $M_1 = (X_1, R_1, v_1)$ and $M_2 = (X_2, R_2, v_2)$, a **bisimulation** between M_1 and M_2 is a binary relation $\sim \subseteq X_1 \times X_2$ such that, whenever $x_1 \sim x_2$, the following conditions are satisfied:

BASE for all $p \in$ PROP, $x_1 \in v_1(p)$ iff $x_2 \in v_2(p)$;
FORTH for all $y_1 \in R_1(x_1)$, there exists a $y_2 \in R_2(x_2)$ with $y_1 \sim y_2$;
BACK for all $y_2 \in R_2(x_2)$, there exists a $y_1 \in R_1(x_1)$ with $y_1 \sim y_2$.

Despite its length, the intuition behind this definition is fairly simple: we want a way of connecting points in one model with points in another that will guarantee that connected points have the same theory. To ensure this, naturally, connected points must agree on which primitive propositions are true—this is the BASE condition. But more than that, they must also agree, more or less, on which worlds are accessible. In particular, if a world x_1 can access y_1 ($x_1 R_1 y_1$), then any world x_2 that x_1 is connected to via the bisimulation ($x_1 \sim x_2$) must be able to access an "equivalent" world y_2 ($x_2 R_2 y_2$). What does it mean to be "equivalent" in the relevant sense? For this we appeal to the bisimulation itself, and demand that y_1 and y_2 be connected ($y_1 \sim y_2$). This is the FORTH condition. The BACK condition is the same thing starting with an edge in the second model and demanding an "equivalent" edge in the first model.

Figure 3.5 shows a simple example of two models and a bisimulation between them. Notice that one world can play "double duty", so to speak, in simulating more than one world from another model.

Exercise 84

Verify that the relation depicted in Figure 3.5 with the dashed lines does indeed satisfy the definition of a bisimulation, then convince yourself with a few examples that worlds connected by the dashed lines have the same theory.

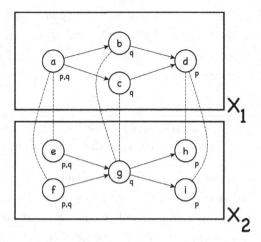

Figure 3.5 A simple bisimulation between two models, depicted by dashed lines.

The definition of a bisimulation has been crafted carefully and intentionally to make true the following theorem, which can be proved by structural induction, analogous to Propositions 42 and 44.

Theorem 45 *If $\sim$ is a bisimulation between M_1 and M_2 and $x_1 \sim x_2$, then $Th_{M_1}(x_1) = Th_{M_2}(x_2)$.*

Exercise 85

Prove Theorem 45.

Exercise 86

Show that Propositions 42 and 44 are consequences of Theorem 45 by defining appropriate bisimulations.

Solution: page 222

Exercise 87

Consider the model $M = (\mathbb{Z}, <, v)$, where $v(p) = \{0\}$ and for all $q \neq p$, $v(q) = \varnothing$. Define a binary relation $\sim$ on $\mathbb{Z}$ by

$$n \sim m \text{ iff } \begin{cases} n, m < 0, \text{ or,} \\ n, m > 0, \text{ or,} \\ n = m = 0. \end{cases}$$

(a) Show that $\sim$ is *not* a bisimulation between M and itself by giving a specific example of how it violates the definition.

(b) Show that the theory of -1 is not the same as the theory of -2. Explain why this also implies that $\sim$ cannot be a bisimulation.

Solution: page 222

As mentioned, invariance results are useful for proving that certain classes of frames *cannot* be defined by any formula in the basic modal language. For example, perhaps somewhat surprisingly, we cannot use this language to pick out all and only the infinite frames.

Proposition 46 *The class of infinite frames,*

$$\mathcal{F}_{inf} = \{(X, R) \; : \; X \text{ is infinite}\},$$

is not defined by any formula in the basic modal language.

Proof Suppose for contradiction that $\varphi \in \mathcal{L}_\square$ defines $\mathcal{F}_{inf}$. Let F be any finite frame; then by definition some model M based on F refutes φ, say at the point x. Now let M' be an infinite model disjoint from M. Consider $M \sqcup M'$; clearly this is an infinite model, so it should validate φ; however, by invariance under disjoint union (Proposition 42), x refutes φ in $M \sqcup M'$ just as it does M. This contradiction shows that φ cannot define $\mathcal{F}_{inf}$.   □

Exercise 88

Show that the following classes of frames are not defined by any formula in the basic modal language.

(a) The class of frames that *contain a self-loop*:

$$\{(X,R) \ : \ \exists x(xRx)\}.$$

(b) The class of frames in which every point has a *predecessor*:

$$\{(X,R) \ : \ \forall x\exists y(yRx)\}.$$

(c) The class of frames that *contain a triangle*:

$$\{(X,R) \ : \ \exists x,y,z(x \neq y, y \neq z, x \neq z, xRy, yRz, \text{and } zRx)\}.$$

(d) The class of *reverse Euclidean* frames:

$$\{(X,R) \ : \ \forall x,y,z((xRz \text{ and } yRz) \text{ implies } xRy)\}.$$

Solution: page 222

Exercise 89

Prove that "being the integers" is not definable in the basic modal language: more precisely, prove that the class of frames $\{(\mathbb{Z}, <)\}$, consisting of the single frame whose worlds are the integers and whose relation is the usual order on $\mathbb{Z}$, is not defined by any formula in the basic modal language.

We can also leverage bisimulations to help us prove frame non-definability results. The key concept is the following: given two frames $F_1 = (X_1, R_1)$ and $F_2 = (X_2, R_2)$, a **bounded morphism** from F_1 to F_2 is a function $f\colon X_1 \to X_2$ that satisfies the FORTH and BACK conditions of a bisimulation.[84]

Theorem 47 *If f is a surjective,[85] bounded morphism from F_1 to F_2, then $F_2 \nvDash \varphi$ implies $F_1 \nvDash \varphi$.*

Proof Since $F_2 \nvDash \varphi$ there is a valuation v_2 such that $M_2 = (F_2, v_2) \nvDash \varphi$. So some $y \in X_2$ is such that $(M_2, y) \nvDash \varphi$. Define $v_1\colon \text{PROP} \to 2^{X_1}$ by $v_1(p) = f^{-1}(v_2(p))$. Then it is easy to check that f is a

---

[84] Recall that every function is also a relation (see Appendix A.1), so this makes sense. The BASE condition cannot be demanded since we are dealing with frames, not models.

[85] That is, $(\forall y \in X_2)(\exists x \in X_1)$ $(f(x) = y)$.

bisimulation between $M_1 = (F_1, v_1)$ and M_2. Since f is surjective, there is some $x \in X_1$ with $f(x) = y$. Then by Theorem 45, since $(M_2, y) \nvDash \varphi$, also $(M_1, x) \nvDash \varphi$, so $F_1 \nvDash \varphi$, as desired. $\qquad\square$

Corollary 48 *If $\mathcal{F}$ is a class of frames, $F_1 \in \mathcal{F}$, and $F_2 \notin \mathcal{F}$, and there exists a surjective, bounded morphism from F_1 to F_2, then $\mathcal{F}$ is not definable by any formula in the basic modal language.*

Proof Suppose for contradiction that φ defines $\mathcal{F}$. Then $F_2 \nvDash \varphi$, so by Theorem 47, $F_1 \nvDash \varphi$, a contradiction. $\qquad\square$

Exercise 90

Let $X = \{x, y, z\}$ and consider the function $f : \mathbb{Z} \to X$ defined by:

$$f(n) = \begin{cases} x & \text{if } (\exists k \in \mathbb{Z})(n = 3k) \\ y & \text{if } (\exists k \in \mathbb{Z})(n = 3k + 1) \\ z & \text{if } (\exists k \in \mathbb{Z})(n = 3k + 2). \end{cases}$$

(a) Prove that f is a surjective bounded morphism from $(\mathbb{Z}, <)$ to the frame $F = (W, R)$, where $R = W \times W$ is the complete relation on W.

(b) Use part (a) to prove that the following frame properties are not defined by any formula in the basic modal language:
- containing no triangles
- containing no cycles
- having more than three elements
- being infinite.

Solution: page 223

Finally, we close this section by observing that we can also use invariance results to show that certain ways of expanding the basic modal language genuinely increase our expressive power. For instance, although we might "extend" the basic modal language with the unary modality $\Diamond\varphi$ defined as follows:

$$x \vDash \Diamond\varphi \text{ iff } (\exists y \in R(x))(y \vDash \varphi),$$

[86] Indeed, we previously defined $\Diamond\varphi$ to be an abbreviation for $\neg\Box\neg\varphi$.

[87] That is, $E\varphi$ is an *unbounded* existential quantifier, ranging over the entire set of worlds, not just the worlds accessible from x.

this does not genuinely increase the expressive power of the language because $\Diamond\varphi$ is equivalent to a formula in the basic modal language: $\vDash \Diamond\varphi \leftrightarrow \neg\Box\neg\varphi$.[86]

On the other hand, we might extend the basic modal langauge with the unary modality $E\varphi$ defined as follows:[87]

$$x \vDash E\varphi \text{ iff } (\exists y \in X)(y \vDash \varphi).$$

In this case we do obtain a genuinely more expressive language, in the following sense.

[88] In this case we sometimes say that E is not *expressible* in the basic modal language.

Proposition 49 $E\varphi$ is not equivalent to any formula in the basic modal language.[88]

Proof Consider two disjoint models M_1 and M_2 where $M_1 \vDash \neg p$ and $M_2 \nvDash \neg p$. Let x be a point in M_1; then clearly $(M_1, x) \nvDash Ep$, since $[\![p]\!]_{M_1} = \varnothing$. However, since $[\![p]\!]_{M_2} \neq \varnothing$, we can see that $(M_1 \sqcup M_2, x) \vDash Ep$. Proposition 42 tells us that $Th_{M_1}(x) = Th_{M_1 \sqcup M_2}(x)$, so it follows that Ep cannot be equivalent to any formula in the basic modal language. $\square$

Exercise 91

(a) Consider the modality ◄ defined by

$$(M, x) \vDash \blacktriangleleft\varphi \text{ iff } (\forall y \in X)(yRx \text{ implies } (M, y) \vDash \varphi).$$

Prove that ◄ is not expressible in the basic modal language.

(b) Consider the modality Θ defined by

$$(M, x) \vDash \Theta\varphi \text{ iff } (\forall y \in X)((M, y) \vDash \varphi \text{ implies } xRy).$$

Prove that Θ is not expressible in the basic modal language.

(c) Consider the *propositional constant* ↻ defined by

$$(M, x) \vDash \circlearrowleft \text{ iff } xRx.$$

Prove that ↻ is not expressible in the basic modal language.

Solution: page 224

3.5 Deduction

A core focus of logic is *deduction*—the idea of mechanically deriving new formulas from previously accepted formulas using some pre-defined rules. A common way to organize this is to begin with some formulas called *axioms*, which are accepted without justification, and then apply *rules of inference*[89] to successively derive more formulas.[90]

An *axiom system* is any collection of axioms and rules of inference taken together. We will consider many different axiom systems in this chapter, beginning with one of the most fundamental, named K, which consists of the following axiom schemes:

(L1) $\varphi \rightarrow (\psi \rightarrow \varphi)$

(L2) $(\varphi \rightarrow (\psi \rightarrow \chi)) \rightarrow ((\varphi \rightarrow \psi) \rightarrow (\varphi \rightarrow \chi))$

(L3) $(\neg\psi \rightarrow \neg\varphi) \rightarrow (\varphi \rightarrow \psi)$

(K) $\Box(\varphi \rightarrow \psi) \rightarrow (\Box\varphi \rightarrow \Box\psi)$.[91]

The rules of inference of K are *modus ponens* (from φ and $\varphi \rightarrow \psi$ infer ψ) and *necessitation*: from φ, infer $\Box\varphi$.[92]

A K-**deduction** is a finite sequence of formulas $\varphi_1, \ldots, \varphi_n$ in $\mathcal{L}_\Box$ such that, for each $1 \leq i \leq n$, either:

(1) φ_i is an axiom, or

(2) there exist $j, k < i$ such that $\varphi_k = \varphi_j \rightarrow \varphi_i$,[93] or

(3) there exists $j < i$ such that $\varphi_i = \Box\varphi_j$.[94]

In this case, the sequence $\varphi_1, \ldots, \varphi_n$ is said to be a K-*deduction of* φ_n.[95] We write $\vdash_K \varphi$ to indicate that there exists a K-deduction of φ.

More generally, if AX is any axiom system, an AX-**deduction** can be defined analogously by reference to the axioms and rules of inference of AX;[96] we write $\vdash_{AX} \varphi$ to indicate that there exists an AX-deduction of φ. Except where otherwise noted, we restrict our attention to axiom systems AX that are *extensions* of K.[97]

Notice that K itself is an extension of the axiom system presented in Chapter 1 for propositional logic. As such, anything

---

[89] Which are also accepted without justification.

[90] The intuitions and much of the formalism here parallel exactly those for deduction in propositional and predicate logic: see Sections 1.4 and 2.3.

[91] Your eyes do not deceive you: the axiom *system* is named K, whereas this particular axiom scheme is denoted (K).

[92] These rules are abbreviated by "MP" and "Nec", respectively.

[93] That is, φ_i follows from two previous lines of the deduction by modus ponens.

[94] That is, φ_i follows from a previous line by necessitation.

[95] It is easy to see (and useful to observe) that if $\varphi_1, \ldots, \varphi_n$ is a deduction, so is any truncation $\varphi_1, \ldots, \varphi_m$, where $m \leq n$. See Lemma 7.

[96] That is: a finite sequence of formulas where each one is either an axiom of AX or follows from previous line(s) by applying a rule of inference from AX.

[97] That is, axioms systems that contain all the axioms and rules of inference of K.

[98] Or any AX extending K.

we could deduce in classical propositional logic we can also deduce using K.[98] In fact, since axiom schemes (and rules of inference) implicitly quantify over arbitrary formulas, any *substitution instance* of a formula that is deducible in classical propositional logic is also deducible in K.[99]

[99] For example, since $p \to p$ is deducible in classical propositional logic (Proposition 4), $\Box p \to \Box p$ is deducible in K, even though the latter is not actually a formula of propositional logic! In fact, the deduction is essentially the same as the one presented in Proposition 4—see Exercise 92.

Exercise 92

Formalize this notion of a substitution instance and prove that every substitution instance of a formula deducible in classical propositional logic is also deducible in K.

Solution: page 224

There are many examples of propositional deductions in Section 1.4; the reader is invited to review those as needed. Here we'll focus on examples of K-deductions that are more than just translated versions of propositional deductions. To make life easier, we henceforth allow ourselves the shortcut of writing any substitution instance of a propositional tautology as a single line in a K-deduction[100]—after all, completeness of propositional logic (Theorem 15) assures us that all tautologies are deducible![101]

[100] Or any AX-deduction, since as noted our default assumption is that AX extends K.

[101] We'll mark these lines of our deductions with "taut", and start using the word "tautology" in the modal context to refer to any substitution instance of a classical propositional tautology.

Proposition 50 $\vdash_K \Box(\varphi \wedge \psi) \to \Box\varphi$.

Proof We need to show there exists a deduction of $\Box(\varphi \wedge \psi) \to \Box\varphi$, so it's enough to provide one.

$$
\begin{aligned}
\varphi_1 &= (\varphi \wedge \psi) \to \varphi && \text{(taut)} \\
\varphi_2 &= \Box((\varphi \wedge \psi) \to \varphi) && \text{(1 Nec)} \\
\varphi_3 &= \Box((\varphi \wedge \psi) \to \varphi) \to (\Box(\varphi \wedge \psi) \to \Box\varphi) && \text{(K)} \\
\varphi_4 &= \Box(\varphi \wedge \psi) \to \Box\varphi && \text{(2,3 MP)} \quad \Box
\end{aligned}
$$

Proposition 51 $\vdash_K (\Box\varphi \wedge \Box\psi) \to \Box(\varphi \wedge \psi)$.

Proof This deduction is a bit more awkward, chiefly owing to the fact that all the basic steps of reasoning must be represented by applications of modus ponens.

$$\begin{array}{rcll}
\varphi_1 & = & \varphi \to (\psi \to (\varphi \wedge \psi)) & \text{(taut)} \\
\varphi_2 & = & \Box(\varphi \to (\psi \to (\varphi \wedge \psi))) & \text{(1 Nec)} \\
\varphi_3 & = & \Box(\varphi \to (\psi \to (\varphi \wedge \psi))) & \\
& & \to (\Box\varphi \to \Box(\psi \to (\varphi \wedge \psi))) & \text{(K)} \\
\varphi_4 & = & \Box\varphi \to \Box(\psi \to (\varphi \wedge \psi)) & \text{(2,3 MP)} \\
\varphi_5 & = & \Box(\psi \to (\varphi \wedge \psi)) \to (\Box\psi \to \Box(\varphi \wedge \psi)) & \text{(K)} \\
\varphi_6 & = & \varphi_4 \to (\varphi_5 \to (\Box\varphi \to (\Box\psi \to \Box(\varphi \wedge \psi)))) & \text{(taut)} \\
\varphi_7 & = & \varphi_5 \to (\Box\varphi \to (\Box\psi \to \Box(\varphi \wedge \psi))) & \text{(4,6 MP)} \\
\varphi_8 & = & \Box\varphi \to (\Box\psi \to \Box(\varphi \wedge \psi)) & \text{(5,7 MP)} \\
\varphi_9 & = & \Box\varphi \to (\Box\psi \to \Box(\varphi \wedge \psi)) & \\
& & \to ((\Box\varphi \wedge \Box\psi) \to \Box(\varphi \wedge \psi)) & \text{(taut)} \\
\varphi_{10} & = & (\Box\varphi \wedge \Box\psi) \to \Box(\varphi \wedge \psi) & \text{(8,9 MP)} \quad \Box
\end{array}$$

Based on these two deductions, it's a good guess that we also have

$$\vdash_K \Box(\varphi \wedge \psi) \leftrightarrow (\Box\varphi \wedge \Box\psi).$$

This is indeed the case, though it's a bit tedious to write out; still, it's useful to go through the process,[102] if only to gain some first-hand experience with how awkward formal deductions can be.

[102] At least once or twice!

Exercise 93

(a) Prove that $\vdash_K \Box(\varphi \wedge \psi) \leftrightarrow (\Box\varphi \wedge \Box\psi)$ by writing out a formal deduction.

(b) Show more generally that

$$\vdash_K \Box(\varphi_1 \wedge \cdots \wedge \varphi_n) \leftrightarrow (\Box\varphi_1 \wedge \cdots \wedge \Box\varphi_n).$$

Exercise 94

(a) Show that if $\vdash_K \varphi \leftrightarrow \psi$, then also $\vdash_K \Box\varphi \leftrightarrow \Box\psi$.

(b) Use part (a) to show that $\vdash_K \Box\varphi \leftrightarrow \neg\Diamond\neg\varphi$.

Before moving on we should take a moment to reflect on the "necessitation" rule of inference and the intuition underlying it. The most important thing to note about this rule is that it is *not*

103 Indeed, necessitation is
strictly weaker than adding
such an axiom.

104 In the two proofs above we
actually only applied
necessitation to instances of
axioms, but in general we
could apply it to any formula
that had been deduced. For
instance, we could extend
the deduction of Proposition
50 with an application of
necessitation to show that
$\vdash_K \Box(\Box(\varphi \land \psi) \to \Box\varphi)$.

105 By contrast, the implication
$\varphi \to \Box\varphi$ is definitely *not*
valid.

106 In a sense made precise by
soundness and
completeness.

107 See Section 1.4 and
especially the discussion
surrounding Theorem 8 for
a review and examples of
this technique.

108 Adopting the notation of
Section 1.4, this might be
written $\{p\} \vdash_K \Box p$.

109 See also sidenote 116.

the same as adding $\varphi \to \Box\varphi$ as an axiom![103] To see why, think about how we used this rule of inference in the proofs above: we were able to add a box only to those formulas that we had previously *deduced*.[104] In other words, necessitation allows us to put a box on φ just in case φ is itself deducible. Thus, a formula like $p \to \Box p$ is not something we can hope to arrive at using necessitation, since p is not deducible.

As we will see in Section 3.6, all the axioms of K are valid, and all the rules of inference preserve validity. In particular, if φ is valid, so is $\Box\varphi$.[105] Since deductions are supposed to "track the truth",[106] this alone is a justification for the necessitation rule of inference: it tracks the fact that validity of φ implies validity of $\Box\varphi$.

This also sheds some light on why we have *not* opted to define the notion of a "deduction from Γ" in this modal context, where Γ is some set of formulas thought of as (temporary) assumptions. In the propositional case this served us very well: using the Deduction Theorem (Theorem 8), we were able to adopt temporary assumptions, prove things from them, and then "import" them as antecedents of material conditionals.[107] But the necessitation rule of inference throws a monkey wrench into this machine: if we treat a temporary assumption like an axiom, we should be able to apply necessitation to it. So, for example, from the assumption p we should be able to deduce $\Box p$.[108] But then a deduction-like theorem cannot hold, since we have already observed that $p \to \Box p$ is not something we should be able to deduce from K.[109]

3.6 Soundness and completeness

The *goal* of deduction is to arrive at *truths*—or, more precisely, validities. It turns out that K is a very good axiom system for the following two reasons.

- Everything deducible from K is valid.
- Everything that is valid is deducible from K.

These properties are called *soundness* and *completeness*, respectively. Soundness says that deductions never "go wrong"—they always lead to validities. Completeness is even more impressive, because it says that deductions are powerful enough to reach *all* the validities.[110]

Theorem 52 *The axiom system* K *is sound with respect to the class of all frames:* $\vdash_K \varphi$ *implies* $\vDash \varphi$.

Proof The idea of the proof is to first show that every axiom is valid, and then show that validity is preserved by modus ponens and necessitation. As we saw in previous soundness proofs (Theorems 10 and 26), a natural way to organize these arguments is by using an inductive proof, specifically, induction on the length of the deduction of φ.

More precisely, we will show that for all n, if $\varphi_1, \ldots, \varphi_n$ is a K-deduction, then $\vDash \varphi_n$. In the base case where $n = 1$, we consider a one-line deduction, which means that φ_1 must be an instance of an axiom scheme.[111] We have already shown that (K) is valid (Proposition 35). The validity of (L1)–(L3) is left as a simple exercise.

Exercise 95

Prove the following.

(a) $\vDash \varphi \rightarrow (\psi \rightarrow \varphi)$
(b) $\vDash (\varphi \rightarrow (\psi \rightarrow \chi)) \rightarrow ((\varphi \rightarrow \psi) \rightarrow (\varphi \rightarrow \chi))$
(c) $\vDash (\neg\psi \rightarrow \neg\varphi) \rightarrow (\varphi \rightarrow \psi)$

This completes the base case. Now suppose the result holds for all deductions of length up to n, and consider a deduction of length $n + 1$: $\varphi_1, \ldots, \varphi_{n+1}$. By the definition of a deduction, we know that φ_{n+1} is either an instance of an axiom, or it follows by modus ponens or necessitation. If it's an axiom we have already shown it must be valid, so we attend to the two rules of inference.

[110] These are essentially the same concepts studied in propositional and predicate logic, interpreted here in the modal setting of frames and models.

[111] So, as promised, we are starting by showing that every axiom is valid.

First suppose that φ_{n+1} follows by modus ponens, so there exist φ_j, φ_k with $j, k < n + 1$ and $\varphi_k = \varphi_j \rightarrow \varphi_{n+1}$. In this case, since $\varphi_1, \ldots, \varphi_j$ and $\varphi_1, \ldots, \varphi_k$ are deductions of length less than $n + 1$, we know by the inductive hypothesis that both φ_j and φ_k are valid. We want to show that φ_{n+1} is valid, so take any model M. Since φ_j is valid, we know in particular that $M \vDash \varphi_j$ so $[\![\varphi_j]\!] = X$. Moreover, since φ_k is valid, we know that $M \vDash \varphi_j \rightarrow \varphi_{n+1}$, so (by Proposition 33) $[\![\varphi_j]\!] \subseteq [\![\varphi_{n+1}]\!]$. Together these imply that $X \subseteq [\![\varphi_{n+1}]\!]$, whence $[\![\varphi_{n+1}]\!] = X$, so $M \vDash \varphi_{n+1}$, as desired.

Now suppose that φ_{n+1} follows by necessitation, so there exists φ_i with $i < n + 1$ and $\varphi_{n+1} = \Box\varphi_i$. Then as before since $\varphi_1, \ldots, \varphi_i$ is a deduction of length less than $n + 1$, we know by the inductive hypothesis that φ_i is valid. Again, we want to show that φ_{n+1} is valid, so let M be an arbitrary model and x a point therein. Then $(M, x) \vDash \varphi_{n+1}$ iff $(M, x) \vDash \Box\varphi_i$ iff $R(x) \subseteq [\![\varphi_i]\!]$, and this last containment is guaranteed to hold since φ_i is valid so $[\![\varphi_i]\!] = X$. $\qquad\qquad\square$

The way that Theorem 52 is phrased hints at a more general formulation: an axiom system may be sound (or complete) with respect to a *class* of frames. Specifically, given an axiom system AX and a class of frames $\mathcal{F}$, we say that AX is **sound** with respect to $\mathcal{F}$ if $\vdash_{AX}\varphi$ implies $\mathcal{F} \vDash \varphi$, and **complete** with respect to $\mathcal{F}$ if $\mathcal{F} \vDash \varphi$ implies $\vdash_{AX}\varphi$.

It is not hard to see, for example, that the axiom system called T, which is obtained by adding all instances of $\Box\varphi \rightarrow \varphi$ as axioms to K,[112] is sound with respect to the class of all reflexive frames. Indeed, the proof of Theorem 52 already gets us most of the way there: it simply needs to be augmented with an argument that each of the newly introduced axioms is valid in $\mathcal{F}_{ref}$, which we have already established (Proposition 38).

A few of the more common[113] additional axioms considered in epistemic contexts are presented below, followed by some of the standard axiom systems built from them.

Soundness results are also useful for showing us which formulas are *not* deducible in a given axiom system. For example,

---

[112] To denote this we write

$$T = K + (\Box\varphi \rightarrow \varphi).$$

[113] And a couple of not-so-common. . .

Table 3.2 A few of the most common (epistemic) modal axioms.

Symbol	Axiom scheme	Epistemic description
(K)	$\Box(\varphi \to \psi) \to (\Box\varphi \to \Box\psi)$	"distribution"
(T)	$\Box\varphi \to \varphi$	"factivity"
(4)	$\Box\varphi \to \Box\Box\varphi$	"positive introspection"
(5)	$\neg\Box\varphi \to \Box\neg\Box\varphi$	"negative introspection"
(D)	$\Box\varphi \to \Diamond\varphi$	"consistency"
(B)	$\varphi \to \Box\Diamond\varphi$	?
(.2)	$\Diamond\Box\varphi \to \Box\Diamond\varphi$	??

Table 3.3 A few of the most common (epistemic) modal axiom systems.

Symbol	Axiom system	Epistemic context
K	described above	pre-epistemic
T	K + (T)	(factive) knowledge
S4	T + (4)	positively introspective knowledge
S4.2	S4 + (.2)	??
S5	S4 + (5)	fully introspective knowledge
KD	K + (D)	(consistent) belief
KD45	KD + (4) + (5)	fully introspective belief

$\Box p \to p$ is not deducible from K since, if it were, by Theorem 52 we would have $\models \Box p \to p$, which we have seen is not so (Proposition 36).

Exercise 96

Prove the following.

(a) $\nvdash_K \Box\varphi \to \Diamond\varphi$
(b) $\nvdash_{KD} \Box\varphi \to \varphi$
(c) $\vdash_T \Box\varphi \to \Diamond\varphi$
(d) $\nvdash_{S4} \Diamond\Box\varphi \to \Box\varphi$
(e) $\vdash_{S5} \varphi \to \Box\Diamond\varphi$

Solution: page 225

Although we begin by proving completeness for K, we will subsequently extend the technique to other axiom systems and other types of completeness results.

[115] So Γ is AX-*inconsistent* if there is some finite subset $\{\varphi_1, \ldots, \varphi_n\} \subseteq \Gamma$ such that $\vdash_{AX} \neg(\varphi_1 \wedge \cdots \wedge \varphi_n)$.

[116] This definition of consistency differs from the one we presented for propositional logic (Section 1.4), where we said that Γ is consistent iff one cannot deduce a contradiction from it. This requires defining a notion of "deduction from Γ", which carries some interpretational difficulties that we commented on at the end of Section 3.5. Note, for example, that $\{p, \neg\square p\}$ intuitively ought to be K-consistent (after all, both formulas can be true at the same world), but necessitation yields an immediate contradiction from this set.

Proving completeness for K is significantly more involved.[114] The technique we will use is called the *canonical model construction*: we will build a gigantic model that simultaneously refutes *every* non-theorem of K. This is sufficient for our proof, since the contrapositive of completeness says that $\nvdash_K \varphi$ implies $\nvDash \varphi$.

How can we build this "canonical" model? Intuitively, we are going to put one world into the model for *every possible way things might be*. The way to do this formally, as in the completeness results for propositional and predicate logic, is through the notion of *maximally consistent sets*.

A set of formulas Γ is called AX-**consistent** if for every finite subset $\{\varphi_1, \ldots, \varphi_n\} \subseteq \Gamma$ we have $\nvdash_{AX} \neg(\varphi_1 \wedge \cdots \wedge \varphi_n)$.[115] Note that this definition has nothing to do with truth or validity—it is cashed out entirely in terms of what can be deduced from the given set of formulas.[116] Note also the *finitary* nature of this definition: inconsistency of Γ is always something that can be demonstrated by a deduction: a finite sequence of formulas ending with $\neg(\varphi_1 \wedge \cdots \wedge \varphi_n)$, for some $\varphi_1, \ldots, \varphi_n \in \Gamma$.

Exercise 97

Show that $\{\varphi_1, \ldots, \varphi_n\}$ is AX-consistent iff $\nvdash_{AX} \neg(\varphi_1 \wedge \cdots \wedge \varphi_n)$.

Exercise 98

Show that this definition of consistency agrees with the one given in Section 1.5 when we restrict our attention to classical propositional logic.

The following lemma will be useful later.

Lemma 53 *If $\Gamma \cup \{\varphi\}$ is AX-inconsistent, there exists a finite subset $\{\varphi_1, \ldots, \varphi_n\} \subseteq \Gamma$ such that $\vdash_{AX} (\varphi_1 \wedge \ldots \wedge \varphi_n) \rightarrow \neg\varphi$.*

Proof Since $\Gamma \cup \{\varphi\}$ is AX-inconsistent, it contains a finite subset such that the negation of its conjunction is AX-deducible.

Thus there exist $\varphi_1, \ldots, \varphi_n \in \Gamma$ such that either $\vdash_{AX} \neg(\varphi_1 \wedge \cdots \wedge \varphi_n)$ or $\vdash_{AX} \neg(\varphi_1 \wedge \cdots \wedge \varphi_n \wedge \varphi)$, depending on whether φ itself is "part of" the inconsistency. However, it is easy to see that

$$\vdash_{AX} \neg(\varphi_1 \wedge \cdots \wedge \varphi_n) \to \neg(\varphi_1 \wedge \cdots \wedge \varphi_n \wedge \varphi);$$

indeed, this is a tautology![117] Thus, in either case, we know that $\vdash_{AX} \neg(\varphi_1 \wedge \cdots \wedge \varphi_n \wedge \varphi)$, which in turn is equivalent to[118]

$$\vdash_{AX} (\varphi_1 \wedge \cdots \wedge \varphi_n) \to \neg\varphi,$$

as desired.   □

A set of formulas Γ is called **maximally AX-consistent** if it is AX-consistent and every proper superset of Γ is *not* AX-consistent.[119]

Proposition 54 *Suppose that Γ is AX-consistent. Then Γ is maximally AX-consistent if and only if for every formula φ, either $\varphi \in \Gamma$ or $\neg\varphi \in \Gamma$.*

Proof First suppose that for all φ, either $\varphi \in \Gamma$ or $\neg\varphi \in \Gamma$. To show that Γ is maximally AX-consistent it suffices to show that $\Gamma' \supsetneq \Gamma$ implies Γ' is not AX-consistent. And indeed, if $\Gamma' \supsetneq \Gamma$, then there is some $\psi \in \Gamma' \setminus \Gamma$, but then, since $\psi \notin \Gamma$, we must have $\neg\psi \in \Gamma$, and so $\psi, \neg\psi \in \Gamma'$, from which it follows that Γ' is not AX-consistent since $\vdash_{AX} \neg(\psi \wedge \neg\psi)$.[120]

Conversely, suppose that Γ' is maximally AX-consistent, and suppose for contradiction that $\varphi \notin \Gamma$ and also $\neg\varphi \notin \Gamma$. By maximality, it must then be that both $\Gamma \cup \{\varphi\}$ and $\Gamma \cup \{\neg\varphi\}$ are AX-inconsistent. By Lemma 53 this means that there is some finite subset $\{\varphi_1, \ldots, \varphi_n\} \subseteq \Gamma$ such that $\vdash_{AX} (\varphi_1 \wedge \cdots \wedge \varphi_n) \to \neg\varphi$, and also some finite subset $\{\psi_1, \ldots, \psi_m\} \subseteq \Gamma$ such that $\vdash_{AX} (\psi_1 \wedge \cdots \wedge \psi_m) \to \neg\neg\varphi$. But this implies that $\vdash_{AX} (\varphi_1 \wedge \cdots \wedge \varphi_n \wedge \psi_1 \wedge \cdots \wedge \psi_m) \to (\neg\varphi \wedge \neg\neg\varphi)$, which in turn implies that $\vdash_{AX} \neg(\varphi_1 \wedge \cdots \wedge \varphi_n \wedge \psi_1 \wedge \cdots \wedge \psi_m)$, contradicting the assumption that Γ is AX-consistent.   □

---

[117] Remember we always assume AX extends K, and from K we can deduce all tautologies.

[118] Again, via a propositional tautology—in fact this is essentially the definition of the material conditional.

[119] In other words, a maximally AX-consistent set is a set of formulas that's "as big as it can be" while still remaining consistent. This is the very same definition we gave in the propositional case (Section 1.5), though as noted the modal notion of consistency is somewhat different.

[120] Since this is a tautology.

Exercise 99

The last sentence in the proof of Proposition 54 contains two claims with no explicit justification. Prove these claims are correct. (Hint: use the fact that from AX we can deduce any tautology.)

In light of Proposition 54, we might think of maximally AX-consistent sets as consisting of a long list of consistent "opinions" about the truth or falsity of *every single formula* in the basic modal language.[121]

[121] We proved essentially the same result in the propositional case: Proposition 12.

After all these definitions and all this work, do we even know whether maximally AX-consistent sets exist? The answer is yes, they do, as follows from the following famous lemma.[122]

[122] Again, compare this to Lemma 13.

Lemma 55 (Lindenbaum's Lemma) *Let Γ be any AX-consistent set of formulas. Then there exists a maximally AX-consistent set $\Gamma' \supseteq \Gamma$.*

Proof Let $\varphi_1, \varphi_2, \ldots$ be an enumeration of all formulas in $\mathcal{L}_\square$. Set $\Gamma_0 = \Gamma$, and inductively define[123]

[123] So we start with Γ and either add φ_1 or not, depending on whether it's consistent. Then similarly with φ_2, and φ_3, and so on.

$$\Gamma_{n+1} = \begin{cases} \Gamma_n \cup \{\varphi_{n+1}\} & \text{if this set is AX-consistent} \\ \Gamma_n & \text{otherwise.} \end{cases}$$

Define $\Gamma' = \bigcup_{n=0}^{\infty} \Gamma_n$. Clearly $\Gamma' \supseteq \Gamma$. Moreover, Γ' must be AX-consistent, since if it weren't some finite subset would already be inconsistent, but that finite subset would be contained in some Γ_n, contradicting the fact that each such set is AX-consistent. Finally, Γ' must be maximally AX-consistent, since for each formula φ, either it is in Γ', or else it was rejected for addition to some Γ_n because $\Gamma_n \cup \{\varphi\}$ is AX-inconsistent, which of course implies that $\Gamma' \cup \{\varphi\}$ is also AX-inconsistent. □

Corollary 56 *A maximally consistent set exists.*

Proof By Lindenbaun's Lemma, since $\varnothing$ is AX-consistent, it can be extended to some maximally AX-consistent set Γ'. □

At long last, we can build our canonical model $\tilde{M} = (\tilde{X}, \tilde{R}, \tilde{v})$.[124]
The set of worlds is given by

$$\tilde{X} = \{x_\Gamma \ : \ \Gamma \text{ is maximally AX-consistent}\}.$$

In other words, there is one world for every maximally AX-consistent set.[125] For any formula φ, define

$$\hat{\varphi} = \{x_\Gamma \in \tilde{X} \ : \ \varphi \in \Gamma\}.$$

So $\hat{\varphi}$ collects the set of worlds in the canonical model that correspond to maximally AX-consistent sets containing φ. Given that our worlds have a *built-in* connection to formulas, it is natural to define the canonical valuation $\tilde{v} \colon \text{PROP} \to 2^{\tilde{X}}$ by

$$\tilde{v}(p) = \hat{p}.$$

So we are saying that p is *true* at exactly those worlds that correspond to maximally AX-consistent sets that contain p. Finally, we define the canonical relation $\tilde{R} \subseteq \tilde{X} \times \tilde{X}$ as follows:

$$x_\Gamma \tilde{R} x_\Delta \text{ iff for all } \varphi, \ \Box\varphi \in \Gamma \text{ implies } \varphi \in \Delta.$$

Intuitively: whenever Γ has the "opinion" that $\Box\varphi$, we are insisting that the only accessible worlds from x_Γ be worlds x_Δ where Δ has the "opinion" that φ.

The main point of all these definitions is to be able to prove the following lemma, which says that the "opinions" of each world in the canonical model match exactly with the formulas that are actually true at those worlds.

Lemma 57 (Truth Lemma) *For all formulas φ, we have* $[\![\varphi]\!]_{\tilde{M}} = \hat{\varphi}$.[126]

Once we've proved this, we can establish the main result!

Corollary 58 K *is complete with respect to the class of all frames:* $\vDash \varphi$ *implies* $\vdash_K \varphi$.

Proof We show the contrapositive: for every φ that is not deducible from K, we have $\nvDash \varphi$. For this it suffices to show that $\tilde{M} \nvDash \varphi$.

[124] In fact we are building a canonical model for each axiom system AX; technically this means we should name it something more like $\tilde{M}_{AX} = (\tilde{X}_{AX}, \tilde{R}_{AX}, \tilde{v}_{AX})$, but this is so notationally cumbersome that we'll suppress the axiom system.

[125] This connects to the intuition, raised earlier, that a maximally AX-consistent set has an "opinion" about every formula—in total, it represents a full description (in the basic modal language) of one way the world might be.

[126] In other words, for all formulas φ and every $x_\Gamma \in \tilde{X}$, we have $(\tilde{M}, x_\Gamma) \vDash \varphi$ iff $\varphi \in \Gamma$.

127 If it weren't, we would have
$\vdash_K \neg\neg\varphi$, which yields a
contradiction—why?

And indeed, if $\nvdash_K \varphi$, then clearly $\{\neg\varphi\}$ is K-consistent,[127] so by Lindenbaum's Lemma (Lemma 55) there is a maximally K-consistent set $\Gamma' \supseteq \{\neg\varphi\}$; but then by the Truth Lemma (Lemma 57), we have $x_{\Gamma'} \models \neg\varphi$, so $\tilde{M} \nvDash \varphi$, as desired. $\square$

So what remains is to prove the Truth Lemma. To do so, it will be useful to first establish the following result.

Lemma 59 (Existence Lemma) *Let Γ be a maximally AX-consistent set, and suppose that $\Diamond\varphi \in \Gamma$. Then there exists a maximally AX-consistent set Δ such that $\varphi \in \Delta$ and $x_\Gamma \tilde{R} x_\Delta$.*

Proof Since we are trying to show that a certain maximally AX-consistent set exists, it makes sense to try to use Lindenbaum's Lemma (Lemma 55), somehow. We want Δ to have two properties: first, to contain φ, and second, to be accessible from Γ.[128] To be accessible from Γ, by definition, Δ must contain the formula ψ whenever Γ contains $\Box\psi$. So let's define the set

$$\Gamma^\Box = \{\psi \; : \; \Box\psi \in \Gamma\},$$

and observe that if we can prove the existence of a maximally AX-consistent set Δ that contains $\Gamma^\Box \cup \{\varphi\}$, we'll be done!

This is where Lindenbaum's Lemma comes in: we'll show that $\Gamma^\Box \cup \{\varphi\}$ is AX-consistent, from which it follows that it can be extended to a maximally AX-consistent set, producing the desired set Δ. So suppose for contradiction that $\Gamma^\Box \cup \{\varphi\}$ is AX-inconsistent; then, by Lemma 53, we know that there is some finite collection of formulas $\psi_1, \ldots, \psi_n \in \Gamma^\Box$ such that

$$\vdash_{AX} (\psi_1 \wedge \cdots \wedge \psi_n) \to \neg\varphi.$$

Then, by necessitation, we must also have

$$\vdash_{AX} \Box((\psi_1 \wedge \cdots \wedge \psi_n) \to \neg\varphi),$$

and so using distribution and modus ponens we can deduce that

$$\vdash_{AX} \Box(\psi_1 \wedge \cdots \wedge \psi_n) \to \Box\neg\varphi. \tag{3.2}$$

One can also show that[129]

$$\vdash_{AX}(\Box\psi_1 \wedge \cdots \wedge \Box\psi_n) \to \Box(\psi_1 \wedge \cdots \wedge \psi_n),$$

so putting this together with (3.2) yields

$$\vdash_{AX}(\Box\psi_1 \wedge \cdots \wedge \Box\psi_n) \to \Box\neg\varphi.$$

This is useful because the formulas $\Box\psi_1, \ldots, \Box\psi_n$ are all in Γ, and so it follows that $\Box\neg\varphi$ must also be in Γ.[130] But, by assumption, Γ also contains $\Diamond\varphi$, that is, $\neg\Box\neg\varphi$, which contradicts AX-consistency of Γ. □

Exercise 100

We used the following fact in the proof of the Existence Lemma: if Γ is maximally AX-consistent, $\chi_1, \ldots, \chi_k \in \Gamma$, and

$$\vdash_{AX}(\chi_1 \wedge \cdots \wedge \chi_n) \to \chi,$$

then $\chi \in \Gamma$. Prove this.

The Existence Lemma basically tells us that when Γ has the "opinion" that $\Diamond\varphi$, that opinion is backed up in the canonical model by the existence of an accessible-from-Γ world that has the opinion φ. This is crucial in the proof of the Truth Lemma, which we can now finally present.

Proof of the Truth Lemma The proof proceeds by induction on φ. For the base case where $\varphi = p \in \text{PROP}$, we have

$$[\![p]\!]_{\tilde{M}} = \tilde{v}(p) = \hat{p},$$

by definition of $\tilde{v}$.

So suppose inductively that the result holds for φ and ψ; we wish to show it holds for $\neg\varphi$, $\varphi \wedge \psi$, and $\Box\varphi$. Observe that

$$x_\Gamma \vDash \neg\varphi \text{ iff } x_\Gamma \nvDash \varphi$$

$$\text{iff } \varphi \notin \Gamma$$

$$\text{iff } \neg\varphi \in \Gamma.$$

The first equivalence is simply the semantic definition of negation, the second is the inductive hypothesis, and the third follows from the fact that Γ is a maximally AX-consistent set, as the reader is invited to check.[131] The inductive step for conjunction is similar.

[131] The forward implication comes from maximality, the backward implication from AX-consistency. Compare this to the corresponding step in the proof of the Truth Lemma for propositional logic (Lemma 14).

Now we attend to the hard part, the inductive step for $\square$. First suppose that $\square\varphi \in \Gamma$; we want to show that $x_\Gamma \vDash \square\varphi$. Let $x_\Delta \in \tilde{R}(x_\Gamma)$; then by definition of $\tilde{R}$, we know that $\varphi \in \Delta$. But then the inductive hypothesis tells us that $x_\Delta \vDash \varphi$; since this holds for arbitrary $x_\Delta \in \tilde{R}(x_\Gamma)$, we have established that $x_\Gamma \vDash \square\varphi$, as desired.

Conversely, suppose that $\square\varphi \notin \Gamma$; we wish to show that $x_\Gamma \nvDash \square\varphi$. By maximality, we must have $\neg\square\varphi \in \Gamma$, and thus $\Diamond\neg\varphi \in \Gamma$.[132] Now by the Existence Lemma, we know there is some $x_\Delta \in \tilde{R}(x_\Gamma)$ with $\neg\varphi \in \Delta$. This implies that $\varphi \notin \Delta$, so by the inductive hypothesis, $x_\Delta \nvDash \varphi$, and therefore $x_\Gamma \nvDash \square\varphi$. $\square$

[132] To see why, compare Exercises 94 and 100.

That's it! That's enough to establish completeness for K with respect to the class of all frames, as shown in Corollary 58.

3.7 Extending completeness results

As with the soundness result, we can extend this proof technique to establish completeness for other axiom systems as well. For example, let's suppose we want to prove that T is not only sound with respect to the class of all reflexive frames, but also complete with respect to this class. What should we do?

We can, once again, build a canonical model. In fact, the entirety of the canonical model construction presented in Section 3.6 was carried out with respect to an arbitrary axiom system AX.[133] So it's a bit misleading to be calling it *the* canonical model—as noted in sidenote 124, we actually built a whole family of canonical models, parameterized by the underlying axiom system AX.

[133] Well, not totally arbitrary: we repeatedly used the assumption that AX extends the deductive system of classical propositional logic, and moreover, to prove the Existence Lemma (Lemma 59), we additionally used the assumption that AX extends K.

What difference does the axiom system make? It determines the notion of consistency, which in turn determines which worlds there are! For example, the canonical model construction for K and for T contain different worlds, because there are sets that are maximally K-consistent but not maximally T-consistent. Indeed, one can prove the following.

Exercise 101

Prove that any set that contains both $\Box p$ and $\neg p$ is T-inconsistent, yet there is a maximally K-consistent set that contains both these formulas.

So the canonical model construction produces a different model for T than it does for K, but all the lemmas we proved were independent of whether AX was equal to K or T or some other system. So are we done? Do we have completeness for T with respect to the class of reflexive frames?

The answer is no. But what is missing?[134]

If we look again at the proof of Corollary 58, all we needed to show was that any formula not K-deducible was refuted on *some* model—it didn't matter which model. But now we are trying to prove that T is complete *with respect to the class of reflexive frames*—and for this we must show that any formula not T-deducible is refuted on some *reflexive* model. In other words, the completeness proof for T doesn't work unless the canonical model for T is reflexive.

Proposition 60 *The canonical model for T is reflexive.*

Proof We need to show that for all x_Γ, $x_\Gamma \tilde{R} x_\Gamma$. By definition of $\tilde{R}$, this is equivalent to showing that whenever $\Box \psi \in \Gamma$, also $\psi \in \Gamma$. This follows from Exercise 100, since $\vdash_T \Box \psi \to \psi$.   □

Corollary 61 *T is complete with respect to the class of reflexive frames.*

This is a fairly general technique for using the canonical model construction to establish completeness of other axiom systems with respect to specific classes of spaces.

[134] Try to figure it out before reading further. Hint: ask yourself where the class of *reflexive* frames came into play.

Exercise 102

(a) Prove that S4 is complete with respect to the class of all frames that are reflexive and transitive.

(b) Let K5 = K + (5). Prove that K5 is complete with respect to the class of Euclidean frames.

(a) Prove that S5 is complete with respect to the class of all frames (X, R) where R is an equivalence relation.

(b) Prove that KD is complete with respect to the class of serial frames.

(c) Prove that KD45 is complete with respect to the class of serial, transitive, and Euclidean frames.

(f) Let K1.1 = K + ($\Diamond \varphi \rightarrow \Box \varphi$). Prove that K1.1 is complete with respect to

$$\mathcal{F}_{par} = \{(X, R) \ : \ \forall x(|R(x)| \leq 1)\}.$$

Solution: page 226

This is not the only way to extend completeness results. We can also use the invariance techniques we studied in Section 3.4 to obtain more powerful or simply different completeness results for familiar axiom systems.

For a very simple example: K is sound and complete with respect to the class of all frames that contain a self-loop, that is, all frames (X, R) that contain at least one $x \in X$ with xRx.[135] How can we show this? We have previously seen that K is sound and complete with respect to the class of *all* frames. Clearly this implies that K is sound with respect to any smaller class of frames. But when we shrink the class of frames, completeness *can* fail—for instance, K is definitely *not* complete with respect to the class of reflexive frames, since there are non-theorems of K that cannot be refuted on this class.[136]

Nonetheless, although shrinking a class of frames *can* ruin a completeness result, interestingly, there are many cases where it does not. This is an important conceptual distinction

---

[135] This isn't a particularly useful thing to know, but it does serve as a good illustration of another way that completeness results can be extended.

[136] For instance, $\Box p \rightarrow p$.

between "soundness and completeness" results and the "frame definability" results we studied previously. The fact that, for example, $\Box p \to p$ defines the class of reflexive frames *implies* that it does not define any other class of frames. By contrast, the fact that T is sound and complete with respect to the class of reflexive frames does *not* mean that it cannot be sound and complete with respect to some other class of frames as well. Indeed, we will see that it is.

Above we claimed that K is complete with respect to the class of frames that contain a self-loop. The proof is trivial.[137] Consider the model $M = (X, R, v)$ where $X = \{a\}$ and $R = \{(a, a)\}$.[138] Clearly M contains a self-loop. Let φ be a non-theorem of K, that is, suppose that $\nvdash_K \varphi$. Let $\tilde{M}$ be the canonical model for K. We know that $\tilde{M} \nvDash \varphi$. By invariance under disjoint union, then, we also have $\tilde{M} \sqcup M \nvDash \varphi$. Clearly $\tilde{M} \sqcup M$ contains a self-loop. Thus, we have shown that every non-theorem of K can be refuted on some model that contains a self-loop, which establishes the desired completeness result.

This particular strengthening of the completeness result for K may not seem especially interesting, but we can do better. For example, a frame is called **connected** if it cannot be written as a disjoint union of two other frames.[139]

Exercise 103

Prove that K is sound and complete with respect to the class of connected frames.

One way of interpreting this result is as follows: we could have started this chapter by defining models in such a way as to *insist* that they are connected—that is, ruling out disconnected models from the outset—and if we had done so, it wouldn't have changed our deductive system in any way. It would still be K. Succinctly: K cannot "tell" whether our models are connected.

[137] Based on the proof of Exercise 88(a), which also involves the property of "containing a self-loop", it's reasonable to guess that the proof involves invariance under disjoint union in some way.

[138] It doesn't matter how v is defined.

[139] Equivalently, a frame (X, R) is **connected** if for all distinct points $x, y \in X$, there is an $R^{\leftrightarrow}$-path from x to y, where $R^{\leftrightarrow}$ denotes the symmetric closure of R. Try to prove this!

> ### Exercise 104
>
> Give an example of a class of frames $\mathcal{F}$ such that K is sound and complete with respect to $\mathcal{F}$ and $|\mathcal{F}| = 1$.

> ### Exercise 105
>
> (a) A frame (X, R) is called **rooted** if there exists a point $r \in X$ (called the *root*) such that, for all $x \in X$, rR^*x. (Recall that R^* denotes the reflexive–transitive closure of R; see Section 3.4.) Show that K is complete with respect to the class of all rooted frames.
> (b) Prove that S5 is complete with respect to the class of all frames (X, R) where $R = X \times X$.
> (c) A frame (X, R) is called a *brush* if there is a non-empty subset $C \subseteq X$ (called the *final cluster* or "handle") such that $R = X \times C$. Prove that KD45 is complete with respect to the class of all brushes.
>
> Solution: page 227

[140] A *cycle* is a finite collection of points $x_1, \ldots, x_n \in X$ such that for all $1 \le i < k$, $x_i R x_{i+1}$, and also $x_n R x_1$. See Appendix A.1.

[141] Intuitively, this forces the frame to have a "branching" structure (hence the designation "tree"), since each R-edge must lead to a new point not reached by any other R-edge.

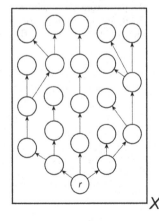

A seemingly central aspect of our models is the fact that the relation is allowed to contain *cycles*.[140] But, as it turns out, we could've ruled that out too without changing the deductive system! In fact, we can push this even farther: we could've assumed that all frames are *trees*.

A frame is called a **tree** if it is rooted with root r and moreover, for every $x \in X$, there is a *unique* R-path from r to x.[141]

> ### Exercise 106
>
> Show that if (X, R) is a tree then it cannot contain any cycles, and for every $x \ne r$, x has a unique predecessor (i.e., there is exactly one y such that yRx).

Proposition 62 K *is sound and complete with respect to the class of trees.*

Proof Soundness is immediate. To establish completeness, we first define a transformation called *unravelling*:[142] the idea is to start with an arbitrary rooted frame (not necessarily a tree), and "unravel" it about the root in a way that produces a tree (with the same root) that can refute the same formulas as the original frame.

How can we define such a transformation? Intuitively, whenever we can get to the same point in two different ways from r, we want to "count" it as a different point. The way to formalize this is to let the *R-paths* in F (starting at r) become the *worlds* of the unravelled frame.

Formally, given a rooted frame $F = (X, R)$ with root r, define

$$\vec{X} = \{(r, x_1, \ldots, x_k) \ : \ (r, x_1, \ldots, x_k) \text{ is an } R\text{-path in } F\},$$

and, given any R-path $\vec{x} = (r, x_1, \ldots, x_k) \in \vec{X}$, set

$$\vec{R}(\vec{x}) = \{\vec{y} \in \vec{X} \ : \ \vec{y} = (r, x_1, \ldots, x_k, y)\}.$$

The frame $\vec{F} = (\vec{X}, \vec{R})$ is called the **unravelling of F about r**.

Also sometimes called *unwinding*.

Exercise 107

Prove that $\vec{F}$ is a tree with root (r) (that is, the root is the unique one-world path $(r) \in \vec{X}$).

Exercise 108

(a) Consider the frame $F = (\{w, a, b\}, \{(w, a), (w, b), (a, b), (b, a)\})$. Draw the tree that is obtained by unravelling F about w.

(b) Consider the frame $F = (\{a, b\}, \{(a, a), (a, b)\})$. What tree is obtained by unravelling F about a?

(c) The *infinite binary tree* is the unique tree such that each node has exactly two children, that is, for all x, $|R(x)| = 2$. Draw a picture of this tree, and find a finite frame that unravels to it.

The intuition here is that we have forced $\vec{F}$ to be a tree by encoding into each point $\vec{x} \in \vec{X}$ the unique $\vec{R}$-path from (r) to $\vec{x}$.[143]

[143] Working through Exercise 107 will make the rest of this proof a lot easier to follow!

Now consider the function $f: \vec{X} \to X$ defined by

$$f(r, x_1, \ldots, x_k) = x_k.$$

So f maps each path in $\vec{X}$ to the world (in F) where it ends. Note that this function is surjective, because F is rooted with root r. In fact, it is also a *bounded morphism*.[144] For the FORTH condition, we need merely observe that $\vec{x}R\vec{y}$ clearly implies $f(\vec{x})Rf(\vec{y})$. For the BACK condition, consider $f(r, x_1, \ldots, x_k) = x_k$ and suppose that $x_k Ry$; then it is easy to see that $(r, x_1, \ldots, x_k)\vec{R}(r, x_1, \ldots, x_k, y)$, and of course $f(r, x_1, \ldots, x_k, y) = y$.

Since bounded morphisms pull back refutations (Theorem 47), what we have shown is that any formula that can be refuted on a rooted frame can also be refuted on a tree! This is a powerful result; indeed, it allows us to finish this proof. Suppose $\nvdash_K \varphi$. Then φ is refuted on some frame, F, say at the point x. Then, by invariance under generated submodels (Proposition 44), we know that φ can also be refuted in (a model based on) the generated subframe F_x. But F_x is rooted![145] So that means that φ can also be refuted on $\vec{F}_x$. Thus we have shown that every non-theorem of K can be refuted on a tree. □

---

[144] As defined in Section 3.4.

[145] Indeed, its root is x.

Corollary 63 K *is sound and complete with respect to the class of frames with no cycles.*

Proof Apply Exercise 106. □

Exercise 109

Given a binary relation R on a set X, the *transitive closure* of R, denoted R^+, is the smallest transitive relation containing R.

(a) Show that xR^+y iff there exists a sequence $z_1, \ldots, z_k \in X$ with $k > 1$ such that $z_1 = x$, $z_k = y$, and for each $i < k$, $z_i Rz_{i+1}$.

(b) Let (X, R) be a rooted frame with root $r \in X$, and let $(\vec{X}, \vec{R})$ denote the unravelling of (X, R) about r. Show that the function $f: \vec{X} \to X$ defined in Proposition 62 might not be a bounded morphism from $(\vec{X}, \vec{R}^+)$ to (X, R), but it is if R itself is transitive.

(c) A *t-tree* is the transitive closure of a tree. Prove that K4 (i.e., K + (4)) is complete with respect to the class of t-trees.

(d) Is K4 complete with respect to the class of finite t-trees? Justify your answer.

Solution: page 228

Exercise 110

Show that K is sound and complete with respect to the class of irreflexive frames:

$$\{(X, R) : \forall x(x \notin R(x))\}.$$

Do *not* use unravelling.

Solution: page 229

The final extension of completeness we will tackle is, arguably, even more impressive and useful. We will develop a technique called *filtration* that we will apply to show that K is complete with respect to the class of *finite* frames.[146]

The idea here is to take an arbitrary model and force it to be finite by "squishing" some of its worlds together. Of course, this kind of squishing will typically change the truth values of some formulas. But we can squish things "carefully" so that some specific formulas of interest have their truth values preserved by the squishing.

We begin with a set of formulas Σ;[147] intuitively, these are the formulas we want to preserve the truth values of. It will be convenient to assume that Σ is **subformula-closed**, which simply means that whenever $\varphi \in \Sigma$ and ψ is a subformula of φ, also $\psi \in \Sigma$.

Next suppose we are given a model $M = (X, R, v)$. For each $x \in X$, define

$$|x|_\Sigma = \{y \in X : (\forall \varphi \in \Sigma)((M, y) \vDash \varphi \text{ iff } (M, x) \vDash \varphi)\}.$$

[146] Which can be used to prove decidability of the deductive system.

[147] Not necessarily maximally consistent, or even consistent!

It is easy to see that the collection $\{|x|_\Sigma \; : \; x \in X\}$ is a *partition* of X.[148] Intuitively, $|x|_\Sigma$ consists of all the worlds in X that agree with x about the truth value of every formula in Σ. This represents our "squishing": we are squishing together all the worlds that agree on the formulas in Σ.

The **filtration of M through** Σ is the model $M_\Sigma = (X_\Sigma, R_\Sigma, v_\Sigma)$ defined as follows:

- $X_\Sigma = \{|x|_\Sigma \; : \; x \in X\}$;
- $|x|_\Sigma R_\Sigma |y|_\Sigma$ iff for some $x' \in |x|_\Sigma$ and $y' \in |y|_\Sigma$ we have $x'Ry'$;
- $v_\Sigma(p) = \begin{cases} \{|x|_\Sigma \; : \; (\forall x' \in |x|_\Sigma)(x' \in v(p))\} & \text{if } p \in \Sigma \\ \varnothing & \text{otherwise.} \end{cases}$

Note that when $p \in \Sigma$, all the worlds in $|x|_\Sigma$ agree on the truth value of p, so the universal quantifier in the definition of v_Σ could just as well have been an existential quantifier.[149]

The key property of M_Σ is that it preserves the truth values of the formulas in Σ.

Proposition 64 *For every φ, if $\varphi \in \Sigma$ then for all $x \in X$, we have $(M, x) \vDash \varphi$ iff $(M_\Sigma, |x|_\Sigma) \vDash \varphi$.*

Proof Induction on φ. For the base case, suppose $\varphi = p \in \text{PROP}$ and $p \in \Sigma$. Then we have

$$(M, x) \vDash p \text{ iff } x \in v(p) \text{ iff } |x|_\Sigma \in v_\Sigma(p) \text{ iff } (M_\Sigma, |x|_\Sigma) \vDash p,$$

as desired.

The inductive steps for negation and conjunction are easy.[150] So suppose inductively that the result holds for φ; we wish to show it holds for $\Box\varphi$. We assume that $\Box\varphi \in \Sigma$, as otherwise this is trivial.

First suppose that $(M, x) \vDash \Box\varphi$. We must show that $(M_\Sigma, |x|_\Sigma) \vDash \Box\varphi$. So let $|y|_\Sigma \in R_\Sigma(|x|_\Sigma)$ be an arbitrary world in M_Σ that is accessible from $|x|_\Sigma$. By definition, this means that there is some $x' \in |x|_\Sigma$ and some $y' \in |y|_\Sigma$ such that $x'Ry'$. Since $x' \in |x|_\Sigma$ and $\Box\varphi \in \Sigma$ we know that $x' \vDash \Box\varphi$. Thus $y' \vDash \varphi$. So, by the inductive hypothesis,[151] we know that $|y'|_\Sigma \vDash \varphi$. Since $y' \in |y|_\Sigma$, we have $|y|_\Sigma = |y'|_\Sigma$, so $|y|_\Sigma \vDash \varphi$, which establishes that $|x|_\Sigma \vDash \Box\varphi$, as desired.

Conversely, suppose that $(M_\Sigma, |x|_\Sigma) \vDash \Box\varphi$. We must show that $(M, x) \vDash \Box\varphi$. So let $y \in R(x)$. Since xRy, we know by definition that $|x|_\Sigma R_\Sigma |y|_\Sigma$, thus $|y|_\Sigma \vDash \varphi$. By the inductive hypothesis, then, we have $y \vDash \varphi$, which establishes that $x \vDash \Box\varphi$, as desired.   □

So we have shown that filtrations preserve the truth values of all formulas in Σ. But what does this have to do with *finite* models? The connection lies in the following fact.

Proposition 65 *If Σ is finite, so is M_Σ. In fact, M_Σ cannot contain more than $2^{|\Sigma|}$ worlds.*

Exercise 111

Prove Proposition 65. (First try it for $\Sigma = \{p\}$ to get a sense of why this is true.)

At last, we are ready to prove the completeness theorem we have been aiming for.

Theorem 66 K *is complete with respect to the class of all finite frames.*

Proof Suppose that $\nvdash_K \varphi$. Then φ is refuted on some model, M. Let Σ_φ denote the set of all subformulas of φ. Clearly, Σ_φ is finite, therefore by Proposition 65, so is M_{Σ_φ}. It is also clear that Σ_φ is subformula-closed, so from Proposition 64 we can deduce that, since $M \nvDash \varphi$, also $M_{\Sigma_\varphi} \nvDash \varphi$. We have therefore shown that φ is refuted on a finite model.   □

Exercise 112

(a) Show that if M is reflexive, so is M_Σ. Deduce that T is complete with respect to the class of finite reflexive frames.

(b) Show that even if M is transitive, M_Σ may not be.

Solution: page 230

Exercise 113

(a) A frame (X, R) is said to have *bounded degree* if there exists a natural number $b \in \mathbb{N}$ such that $(\forall x \in X)(|R(x)| \leq b)$. Show that K is complete with respect to the class of trees that have bounded degree.

(b) Show that every non-theorem of KD can be refuted at a point of infinite degree (i.e., a point x with $|R(x)| = \infty$), but the same is not true for K.

Solution: page 230

4 Group Knowledge

Reasoning about knowledge makes sense in both an individual and a group context. Indeed, the very first motivating example we presented in detail in Section 3.1—the hat puzzle—was a group context where we specifically cared about what the players knew about each other, what they knew about what others knew, and so on.

In this short "exploration" chapter we expand the formal framework presented in Chapter 3 so as to incorporate multiple agents. While this basic expansion is fairly simple, it opens the way to defining further extensions of the logic to capture group-specific concepts such as *mutual knowledge* and a stronger notion that has traditionally been called *common knowledge*.

4.1 Multi-agent epistemic logic

We begin, unsurprisingly, by adjusting the basic modal language to incorporate multiple "knowledge" modalities, one for each agent we wish to reason about. The *n*-**agent epistemic language** is defined recursively as follows:

$$\varphi ::= p \mid \neg\varphi \mid \varphi \wedge \psi \mid K_i\varphi,$$

where $p \in \text{PROP}$, a countable set of primitive propositions (as usual), and $i \in \{1, \ldots, n\}$, representing the different agents. Let $\mathcal{L}_K^n$ denote the set of all formulas in this language. For convenience we'll write $\hat{K}_i\varphi$ as an abbreviation for the dual of K_i, namely $\neg K_i \neg \varphi$.[1]

[1] The "hat" symbol is meant to be evocative of the "diamond" shape we previously used for the dual.

We define semantics for this language in the most obvious possible way: by providing an accessibility relation for each agent.[2] Since we do not wish to hardcode any necessary connection between what different agents know, this makes our job even simpler: these n accessibility relations are completely independent of one another.

Formally, an n-**frame** is a tuple $(X, R_1, \ldots, R_n)$ where X is a non-empty set and each R_i is a binary accessibility reluation on X. An n-**model** is simply an n-frame together with a valuation $v \colon \mathrm{PROP} \to 2^X$.[3] We will further assume in this chapter that each relation is at least reflexive, since this naturally corresponds to factivity of each agent's knowledge modality—a fairly uncontroversial assumption.

[3] We'll frequently drop the n-prefix when it's clear from context.

The truth set of an arbitrary formula in $\mathcal{L}_K^n$ in a model $M = (X, R_1, \ldots, R_n, v)$ is defined recursively, as usual:

$$[\![p]\!]_M = v(p)$$
$$[\![\neg\varphi]\!]_M = X \setminus [\![\varphi]\!]_M$$
$$[\![\varphi \wedge \psi]\!]_M = [\![\varphi]\!]_M \cap [\![\psi]\!]_M$$
$$[\![K_i\varphi]\!]_M = \{x \in X : R_i(x) \subseteq [\![\varphi]\!]_M\}.$$

So far we have done nothing impressive—just "copied" n times the syntactic and semantic structures we saw in Chapter 3. Nonetheless, this expanded language can already express the kind of complicated nested knowledge statements that were so important in the hat puzzle, for instance, as follows.

[4] An important subtlety when reading negated knowledge modalities is that English expressions like "j doesn't know φ" sometimes feel like they carry the implication that φ really is true (j just doesn't realize it). Here this is not the case: $x \vDash \neg K_j\varphi$ exactly when there is some $y \in R_j(x)$ such that $y \vDash \neg\varphi$. And this can be the case whether $x \vDash \varphi$ or not. In particular, "j doesn't know that i knows p" does *not* imply that i really does know p.

- K_iK_jp: "Agent i knows that j knows that p."
- $\neg K_iK_jK_ip$: "Agent i doesn't know that j knows that i knows p."
- $\hat{K}_i(K_jp \wedge \neg K_jK_ip)$: "Agent i considers it possible that j knows p but that j doesn't know that i knows p."[4]

After presenting and analyzing the hat puzzle (Section 3.1), we left the reader with an additional intuitive puzzle: Exercise 63 asked how it could be the case that telling a group of people something they all already knew could make a difference to their reasoning, referring specifically to the statement, "At least one of you is wearing a red hat."

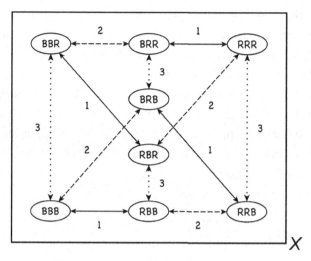

Figure 4.1 A 3-model of the hat puzzle
(all reflexive edges are suppressed).

We can now formalize some of the key concepts in this
question. Consider the 3-model depicted in Figure 4.1 (adapted
slightly from Figure 3.1). Let $red_A, red_B, red_C \in$ PROP be primitive
propositions standing for Alice, Bob, or Carl wearing a red hat,
respectively.

$$v(red_A) = \{\text{RRR, RBR, RRB, RBB}\}$$

$$v(red_B) = \{\text{RRR, BRR, RRB, BRB}\}$$

$$v(red_C) = \{\text{RRR, BRR, RBR, BBR}\}$$

Then the formula $\alpha = red_A \lor red_B \lor red_C$ coincides with the
statement of interest, since it holds at all and only those worlds
where at least one player is wearing a red hat.[5] And we can see
directly that at world RRR, which in the story is the real world,
each player does know α:[6]

$$\text{RRR} \vDash K_1\alpha \land K_2\alpha \land K_3\alpha.$$

This is called *mutual knowledge* and it is important enough
to be abbreviated: in general we'll write $E\varphi$ as an abbreviation
for $K_1\varphi \land \cdots \land K_n\varphi$.[7] From a certain logical point of view it is
not very interesting, since, being an abbreviation, it doesn't

---

[5] Indeed, $[\![\alpha]\!] = X \setminus \{\text{BBB}\}$.

[6] For example, player 2 knows
α at RRR because
$R_2(\text{RRR}) = \{\text{RRR, RBR}\} \subseteq [\![\alpha]\!]$.

[7] The letter E here stands for
"everyone knows".

actually allow us to express anything we couldn't already say in the standard n-agent epistemic language. Nonetheless, as we'll see, it is an important conceptual stepping stone to much more.

We can actually think of E not just as an abbreviation, but also a "box-type" modality in its own right. This is because, by definition, $E\varphi$ holds at x if and only if for all i, $R_i(x) \subseteq [\![\varphi]\!]$, and this condition in turn can be captured by (bounded) universal quantification over a single notion of accessibility, namely, that given by the relation $R = R_1 \cup \cdots \cup R_n$. In other words, if we define a new relation R on X by setting

$$xRy \text{ iff for some } i, xR_iy,$$

then it is easy to check that $x \vDash E\varphi$ iff $R(x) \subseteq [\![\varphi]\!]$.

Exercise 114

Prove that if $R_1, \ldots, R_n$ are reflexive, so is their union, R. Show by contrast that even if $R_1, \ldots, R_n$ are transitive, R may be fail to be transitive, and a similar result holds for the Euclidean property.

Exercise 114 is a relatively simple graph-theoretic problem exploring which properties of relations are preserved by unions. But given the correspondence between epistemic and graph-theoretic properties established in Chapter 3, we can also view the results of this exercise from an epistemic perspective.

The preservation of reflexivity tells us that if "knowing" is factive, so is "everybody knowing"; in other words, if individuals can't know false things, then groups also cannot mutually know false things. This seems quite intuitive. On the other hand, the failure of transitivity tells us that even if we begin with positively introspective agents—that is, agents who know what they know—that kind of introspection won't necessarily hold at the group level. Again, this seems quite intuitive: just because everyone knows φ, that doesn't mean that everyone know that everyone knows φ! Notice that this latter claim requires not only personal introspection, but also for each member of the group to know that each *other* member knows φ. For example, when $n = 2$, the formula $EE\varphi$ is an abbreviation for

$$K_1(K_1\varphi \wedge K_2\varphi) \wedge K_2(K_1\varphi \wedge K_2\varphi),$$

which of course is equivalent to

$$K_1K_1\varphi \wedge K_1K_2\varphi \wedge K_2K_1\varphi \wedge K_2K_2\varphi.$$

So agent 1 has to know that agent 2 knows φ, and agent 2 has to know that agent 1 knows φ. This kind of interpersonal knowledge does not follow from mutual knowledge, even for introspective agents.[8]

This line of reasoning offers us a clue to the original challenge: how can α be informative when announced to a group for whom α is already mutually known? Well, we have just seen that it is possible for φ to be mutually known, yet $E\varphi$ not to be,[9] so perhaps in the hat puzzle this is the information that's gained: everybody already *individually* knew α, but when it was announced they learned that not only is α true, but also that everyone knows it now.

This is in the right spirit of the solution to the puzzle, but not quite there yet: looking closely at Figure 4.1, we can actually see that RRR $\models EE\alpha$ already! However, RRR $\not\models EEE\alpha$. In other words: everyone knows that there's at least one red hat, and everyone knows that everyone knows that,[10] but they don't all know that they all know that they all know that. For example, Alice considers it possible that Bob considers it possible that Carl considers it possible that α is false:

RRRR_1BRR, BRRR_2BBR, BBRR_3BBB, and BBB $\not\models \alpha$.

The core idea here is that when information is announced to a group, the result is not just that each person learns the thing that is announced, but they also learn that they have all learned it, and that they have all learned that they have all learned it, and so on, ad infinitum. This may seem perverse, but we have just seen an example that relies precisely on the distinction between $EE\varphi$ and $EEE\varphi$.

[8] Try for yourself to translate the failure of the Euclidean property for E into epistemic terms.

[9] That is, $E\varphi \wedge \neg EE\varphi$.

[10] After all, each person is seeing two red hats, so they know the other two are seeing at least one.

Exercise 115

Draw a model for a four-player version of the hat puzzle; show that at the world RRRR, we have $E\alpha \wedge EE\alpha \wedge EEE\alpha$ but not $EEEE\alpha$, where $[\![\alpha]\!] = X \setminus \{BBBB\}$.

These considerations highlight the conceptual importance of iterated mutual knowledge, and demonstrate that there is no fixed upper bound on how many iterations are relevant for reasoning. Indeed, it seems that even in the relatively mundane context of announcing something to a group, the information that is thereby imparted to the members of that group can include arbitrarily deeply nested mutual knowledge.

In light of this, it will be convenient to introduce new abbreviations for iterated mutual knowledge: recursively define $E^0\varphi$ to be φ and, for $k \in \mathbb{N}$, let $E^{k+1}\varphi$ abbreviate $EE^k\varphi$.[11] In this notation, the concept we have arrived at can be expressed, at least informally, as

$$\bigwedge_{k=1}^{\infty} E^k\varphi. \tag{4.1}$$

This has a technical name: it is called *common knowledge*.

4.2 Common knowledge

The expression in (4.1) is not actually a formula in our language—it's infinitely long, and the recursive definition of $\mathcal{L}_K^n$, as with every other language considered in this book, can only ever produce finite formulas.[12]

While this may seem like a substantial obstacle to reasoning formally about common knowledge, somewhat surprisingly, we can bypass it fairly easily.[13] Let's introduce a new (unary) modality into our language—call it C, for "common knowledge"—and denote the resulting expanded language by $\mathcal{L}_{CK}^n$. As we've highlighted, what we *cannot* do is declare that $C\varphi$ is an abbreviation for $E\varphi \wedge E^2\varphi \wedge E^3\varphi \wedge \cdots$, precisely because this is not a formula. But what we *can* do is extend our semantic

[11] So the upshot is that E^k abbreviates writing out the "E" modality k times.

[12] There are such things as *infinitary logics*, which can permit expressions like that in (4.1) as bona fide formulas, but the study of these logics lies beyond the scope of this book.

[13] Though it will create some interesting difficulties down the road.

clauses to directly define the truth conditions for $C\varphi$ in n-models. In other words, we can treat C like a brand new modality.[14]

[14] Similar to how we introduced additional knowledge modalities at the beginning of this chapter.

Of course, we don't want C to be completely independent of the existing knowledge modalities; quite the contrary, the truth of $C\varphi$ should be completely determined by lower-level facts about what's known. How can we arrange this semantically? When should we say that $C\varphi$ is true at x, in a model M? Well, if we want $C\varphi$ to be true exactly when $E^k\varphi$ is true for all k, nothing stops us from simply setting

$$[\![C\varphi]\!]_M = \bigcap_{k=1}^{\infty}[\![E^k\varphi]\!]_M.$$

Thus $(M,x) \vDash C\varphi$ iff for all $k \geq 1$, $(M,x) \vDash E^k\varphi$.

This may seem a bit like cheating, since we are still using an infinite expression, namely the intersection over all k from 1 to ∞. Crucially, though, this expression lies in the metalanguage, not the object language: we are essentially defining a new, finite formula $C\varphi$ in such a way that it has the same semantic value as the infinitary expression in (4.1). As foreshadowed, this will eventually lead to some strange consequences, but for the moment it allows us to reason about the concept of common knowledge alongside knowledge using the familiar logical apparatus we have developed. For instance, we have the following.

Proposition 67 *The following are valid in all n-models.*

(a) $C(\varphi \wedge \psi) \leftrightarrow (C\varphi \wedge C\psi)$
(b) $C\varphi \rightarrow \varphi$
(c) $C\varphi \rightarrow E\varphi$
(d) $C\varphi \rightarrow EC\varphi$
(e) $C\varphi \rightarrow CC\varphi$

In working through Proposition 67, one is likely to uncover the following extremely useful fact which characterizes C, like E and the K_i, as yet another "box-type" modality.

Lemma 68 *Let* $M = (X, R_1, \ldots, R_n, v)$ *be an n-model, let* $R = \bigcup_{i=1}^{n} R_n$, *and let* R^* *be the reflexive–transitive closure of* R.[15] *Then for all* $x \in X$, *we have*

$$(M, x) \vDash C\varphi \text{ iff } R^*(x) \subseteq \llbracket \varphi \rrbracket_M.$$

Proof We must show that C operates by (bounded) universal quantification over R-reachable points; more precisely, that $x \vDash C\varphi$ iff for every y that is R-reachable from x, we have $y \vDash \varphi$. This follows immediately from the definition of C together with the fact that for all $k \geq 1$, for all $x \in X$, $x \vDash E^k \varphi$ iff every R-path $x_1, x_2, \ldots, x_{k+1}$ with $x = x_1$ is such that $x_{k+1} \vDash \varphi$. We can prove this latter fact by induction on k.

For the base case $k = 1$, we must show that $x \vDash E\varphi$ iff whenever $x R x_2$ we have $x_2 \vDash \varphi$, which is immediate. Now suppose inductively the result holds for k. We know that $x \vDash E^{k+1} \varphi$ iff $x \vDash E E^k \varphi$ iff for all $y \in R(x)$, $y \vDash E^k \varphi$. The inductive hypothesis tells us that $y \vDash E^k \varphi$ iff every R-path $y_1, \ldots, y_{k+1}$ with $y = y_1$ is such that $y_{k+1} \vDash \varphi$. To complete the proof, then, we need simply observe that if $y_1, \ldots, y_{k+1}$ is an R-path with $y = y_1$, then $x, y, y_2, \ldots, y_{k+1}$ is also an R-path, and conversely, if $x_1, x_2, \ldots, x_{k+2}$ is an R-path with $x = x_1$, then $x_2 \in R(x)$ and $x_2, \ldots, x_{k+2}$ is an R-path. $\square$

Exercise 116

Prove Proposition 67.

It is easy enough to axiomatize n-agent epistemic logic *without* common knowledge: since all the knowledge modalities are independent of one another, we can simply adapt the system T by expanding (K) and (T) in the obvious way:

(K_i) $K_i(\varphi \to \psi) \to (K_i\varphi \to K_i\psi)$,
(T_i) $K_i\varphi \to \varphi$.

Similarly, we expand the "necessitation" rule of inference to include each modality: from φ infer $K_i\varphi$. Call this system T_n. It is straightforward to adapt the arguments of Section 3.6 to prove soundness and completeness.

Exercise 117

Prove that T_n is a sound and complete axiomatization of $\mathcal{L}_K^n$ with respect to the class of all n-frames. Define $S4_n$ and $S5_n$ in the obvious ways and prove similar results for these systems.

Axiomatizing the richer language $\mathcal{L}_{CK}^n$ is more of challenge. For one thing, translating the validities of Proposition 67 into axioms is not enough to achieve completeness. Intuitively, an axiom like $C\varphi \rightarrow E\varphi$ encodes the fact that common knowledge is stronger than mutual knowledge, but not *how much* stronger. Even if we added as axioms all implications of the form $C\varphi \rightarrow E^k\varphi$, for $k \geq 1$, this would only encode a one-sided bound on the strength of $C\varphi$—namely, the fact it is $C\varphi$ is stronger than arbitrarily deeply nested mutual knowledge. Still missing from the picture is the fact that $C\varphi$ is in fact *equal in strength* to the conjunction of all these $E^k\varphi$.

This would seem to require an axiom in which $C\varphi$ is the consequent of an implication rather than the antecedent, that is, something that says "this (the antecedent) is enough to guarantee common knowledge". What could play that role? After all, none of the $E^k\varphi$ are individually strong enough to imply $C\varphi$, and their infinite conjunction is not a legal antecedent.

One clever way of encoding the true strength of $C\varphi$ without resorting to infinitary formulas is through a new rule of inference called the *induction rule*: from $\psi \rightarrow E(\varphi \wedge \psi)$, infer $\psi \rightarrow C\varphi$. This rule results in an implication with $C\varphi$ in the consequent, like we wanted. But what is this formula ψ that implies $C\varphi$? It is a formula for which the implication $\psi \rightarrow E(\varphi \wedge \psi)$ can be deduced. Perhaps the best intuition for such a ψ is to imagine that ψ is an "announcement of φ" to the group of agents.[16] In this case, roughly speaking, ψ is the kind of thing that implies

[16] Like the announcement at the beginning of the hat puzzle that there is at least one red hat.

[17] Since that's what's being announced.

not only that everyone knows φ,[17] but also that everyone knows ψ itself—namely, the fact that the announcement was made also becomes mutual knowledge, once it's made!

Formally speaking, as with any candidate rule of inference, its justification relies on the fact that is preserves validity.

Proposition 69 $\vDash \psi \to E(\varphi \wedge \psi)$ *implies* $\vDash \psi \to C\varphi$.

Proof Suppose that $\vDash \psi \to E(\varphi \wedge \psi)$ and let $x \vDash \psi$. We must show that $x \vDash C\varphi$. We will first prove that all points y that are R-reachable from x satisfy $\varphi \wedge \psi$ by induction on the length k of the R-path from x to y. For $k = 1$, since $x \vDash E(\varphi \wedge \psi)$, we know that for each $y \in R(x)$ we have $y \vDash \varphi \wedge \psi$.

Now suppose inductively that if z is R-reachable from x via a path of length k, then $z \vDash \varphi \wedge \psi$, and let y be R-reachable from x via a path of length $k + 1$. Let z be the penultimate point on this path; then z is R-reachable from x via a path of length k, so by the inductive hypothesis we know that $z \vDash \varphi \wedge \psi$. We also know that zRy, and since $\vDash \varphi \to E(\varphi \wedge \psi)$, we have $z \vDash E(\varphi \wedge \psi)$. Thus, $y \vDash \varphi \wedge \psi$, which completes the induction.

We have therefore shown that every y that is R-reachable from x satisfies $y \vDash \varphi$. It follows from Lemma 68 that $x \vDash C\varphi$. $\square$

Let $\mathsf{T_n C}$ denote the axiom system obtained from $\mathsf{T_n}$ by adding the axioms

(C1) $C\varphi \to E\varphi$
(C2) $C\varphi \to EC\varphi$

and the induction rule of inference.

Theorem 70 $\mathsf{T_n C}$ *is a sound axiomatization of* $\mathcal{L}_{KC}^n$ *with respect to the class of all n-frames.*

Proof As usual for soundness proofs, this can be established by induction on the length of the deduction, making crucial use of Propositions 67 and 69. $\square$

Exercise 118

(a) Show that $\vdash_{\mathsf{T}_n\mathsf{C}} C\varphi \to CC\varphi$.

(b) Show that for all $k \geq 1$, $\vdash_{\mathsf{T}_n\mathsf{C}} C\varphi \to E^k\varphi$.

Solution: page 231

Completeness is not such a simple adaptation: the canonical model method cannot be straightforwardly applied here, owing essentially to the infinitary nature of common knowledge. For consider the following set of formulas:

$$\Sigma = \{Ep, E^2p, E^3p, \ldots\} \cup \{\neg Cp\}.$$

On the one hand, it is obvious that Σ is not satisfiable—by definition, the infinite collection $\{E^kp : k \geq 1\}$ can only all be true at a world where Cp holds. However, Σ *is* consistent. This is because consistency is a *finitary* notion—to prove Σ inconsistent would be to prove that some finite subset is inconsistent. But it's easy to check that no finite subset of Σ is inconsistent.[18]

So the connection between consistency and satisfiability breaks down for this logic.[19] But how exactly is this a problem for the canonical model method? The set Σ, being consistent, can be extended to some maximal consistent set Σ', which will then become a point in the canonical model for $\mathsf{T}_n\mathsf{C}$. But then the Truth Lemma cannot be true, for it would assert that each formula in Σ' is true at that point in the canonical model, and we have already seen that Σ is unsatisfiable.

With that being said, it turns out that $\mathsf{T}_n\mathsf{C}$ *is* complete with respect to the class of n-frames. Adapting the canonical model method to prove this is an intriguing challenge.

[18] Hint: construct a model where E^kp holds for all k up to some fixed number, but Cp fails. Then appeal to soundness.

[19] In other words, the logic is not *compact*.

5   Topological Semantics

[1] Whatever that means.

[2] These shapes are "topologically equivalent".

[3] If you haven't seen topology before, don't worry—this is designed to be accessible to a newcomer. If you have seen topology before, you probably haven't seen it introduced in quite this way. So there's something new here for everyone.

[4] We adopt the convention of using a solid line to indicate a set that contains its boundary. So y really is in A.

Topology is the mathematical study of space without distance,[1] what is sometimes called "rubber sheet geometry". Perhaps the most canonical image associated with the study of topology is that of a coffee cup deforming into a doughnut (and vice versa).[2] What could this possibly have to do with modal logic?

In this short "exploration" chapter we provide a somewhat idiosyncratic (though complete and rigorous) introduction to topology, and develop the connection to modal logic along the way. Perhaps the strongest intuitive connection surfaces through an *epistemic* interpretation of modality, so we will favour that metaphor as we go.[3]

5.1 Topology

One way to approach topology is through a dissatisfaction with the concept of set membership. Set membership is a binary affair: given $x \in X$ and $A \subseteq X$, either $x \in A$ or $x \notin A$—there is no middle ground.

However, drawing on very basic spatial intuitions, we can look at a diagram like the one in Figure 5.1 and see that x is "more" in A than y is.[4] We might say x is "fully" or "robustly" in A, whereas y is just "barely" in A.

How can we make sense of this? In other words, how could we (minimally) augment the basic structure of sets to represent this distinction between mere membership and "robust" membership?

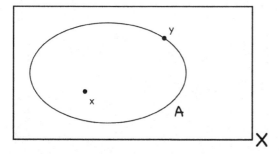

Figure 5.1 Set membership is binary; spatial intuitions are not.

One idea that immediately comes to mind is to incorporate a notion of *distance*. We can see that y, unlike x, is arbitrarily close to elements of $X \setminus A$. So if we added a distance function $d \colon X \times X \to [0, \infty)$, we could try to use it to capture this idea.

The topological approach is actually more subtle than that. Specifying a quantitative distance function[5] is overkill. We can represent spatial structure, and in particular capture a notion of "nearness", in purely qualitative terms—without a numeric distance function. This is what topology does.

[5] What is often called a *metric*.

Exercise 119

Before reading further, explore some of your own ideas about how one might capture the distinction between points like x and y in Figure 5.1 *without* referring to a quantitative notion of distance.

One intuition that might occur to us, if we stare at Figure 5.1 for a while, is that if we could pick up the set X and "shake" it a little, leaving everything just slightly displaced from its original location, then x would end up still being in A, whereas y might not. The same intuition can also be approached not with a "shaking" metaphor, but by thinking in terms of measurement and error.[6] If we were to try to measure the location of x, using (of course) our imperfect measurement tools, we would not get an exact, perfectly precise location, but rather a range of possible locations (represented by a set of points).

[6] This is the first point of contact with epistemology.

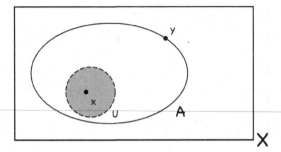

Figure 5.2 A witness to x's membership in A.

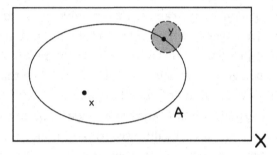

Figure 5.3 Trying to measure y's location.

As long as our measurement is precise enough, we can use it ascertain that x is indeed in A; see Figure 5.2. By contrast, intuitively, no measurement, no matter how precise, seems capable of telling us for sure that y is in A. See Figure 5.3.

This leads to the following first attempt to define the notion of "robust" membership: say that x is *robustly in A* if there exists a "witnessing" set U such that $x \in U$ and $U \subseteq A$. In other words, not only is x a member of A, but there is a kind of "cushion", in the form of the set U, that witnesses x's membership in A.

Exercise 120

Once again, it's worth pausing here to ask yourself whether this is a reasonable definition. In fact, it's not. Try to articulate what's wrong with it.

For one thing, we can always just take $U = \{x\}$, in which case we certainly have $x \in U$, and $U \subseteq A$ just in case $x \in A$. So in fact this definition of robust membership collapses just to plain old membership.

One might object that choosing a singleton set is "cheating", and so modify the definition to rule out singletons. But that doesn't really solve our problem: we can also take $U = A$, in which case clearly $U \subseteq A$, and once again whenever $x \in A$ we have $x \in U$. So again we are left with a definition that doesn't actually make robust membership any different from regular membership.

The solution, and the core idea of topology, is to include the collection of all possible "witnessing" sets U as part of the fundamental structure of the space. That is, a **topological space** is a pair $(X, \mathcal{T})$ where X is a set, and $\mathcal{T} \subseteq 2^X$ is a collection of subsets of X satisfying certain properties.[7] The collection $\mathcal{T}$ is called a *topology* on X, and its elements are called *open sets*.

Intuitively, open sets are precisely the sets that can act as "witnesses" for robust membership. This is formalized in the following definition: a point x is said to be in the **interior** of a set A if there exists an open set $U \in \mathcal{T}$ such that $x \in U \subseteq A$. In this case we write $x \in int(A)$, so $int(A)$ denotes the *interior* of A—the set of all points in the interior of A.

So what we had been calling "robust membership" corresponds to the topological notion of interior, with the crucial addition of the topology, which encodes spatial structure by specifying which sets can act as witnesses.

If we return to the epistemic metaphor of measurement and error, the open sets correspond exactly to the *results of possible measurements or observations*, and, as such, $x \in int(A)$ can be interpreted as saying that there exists some measurement or observation that entails A, at x. Looking again at Figure 5.1, we see that while A is true at both x and y,[8] A is only *measurably* true at x.

It is easy to check that for all A, $int(A) \subseteq A$; that is, the interior function $int: 2^X \to 2^X$ is a *shrinking operator*—it maps every set to

[7] We'll come back to these properties later; for now they are a distraction.

[8] Saying "A is true at x" is just another way of saying that $x \in A$, borrowing terminology (and intuition) from possible world semantics.

a subset of itself. This corresponds to the obvious fact that any-thing that's measurably true is also just true. It's also not hard to show that $int(int(A)) = int(A)$; that is, the interior operator is *idempotent*—applying it a second time does nothing.[9]

[9] It's a good idea to try to prove these facts yourself before reading the proof below.

Proposition 71 *For all A, $int(A) \subseteq A$ and $int(int(A)) = int(A)$.*

Proof First we'll show that $int(A) \subseteq A$. Let $x \in int(A)$. Then by definition there is a $U \in \mathcal{T}$ such that $x \in U$ and $U \subseteq A$, from which it follows immediately that $x \in A$, as desired.

[10] The reverse containment is an immediate consequence of the previous result!

Now we'll show that $int(A) \subseteq int(int(A))$.[10] Let $x \in int(A)$. Once again, by definition this means that there is a $U \in \mathcal{T}$ such that $x \in U \subseteq A$. We'll show that in fact $U \subseteq int(A)$, which will then yield (by definition) that $x \in int(int(A))$, as desired. So let $y \in U$. Then we have $y \in U \subseteq A$, so $y \in int(A)$; since y was arbitrary, this shows that $U \subseteq int(A)$ and completes the proof. $\square$

Exercise 121

Show that for all $A, B \subseteq X$, if $A \subseteq B$, then $int(A) \subseteq int(B)$.

---

[11] So for all $x \in X$, there is at least one $U \in \mathcal{T}$ with $x \in U$.

[12] $\mathcal{T}$ is closed under pairwise intersections.

[13] $\mathcal{T}$ is closed under arbitrary unions.

[14] Closure under pairwise intersection makes sense, from the epistemic perspective—it essentially corresponds to the idea that if we take two measurements or make two observations, we can "combine" what we learned from each one. Closure under arbitrary union, on the other hand, is a bit harder to interpret epistemologically. We'll return to this conceptual point later.

The interior operator has other properties as well, but to prove them we need the full definition of a topological space: a **topological space** is a set X together with a collection $\mathcal{T} \subseteq 2^X$ such that

(T1) $\mathcal{T}$ covers X;[11]
(T2) $U, V \in \mathcal{T}$ implies $U \cap V \in \mathcal{T}$;[12]
(T3) $\mathcal{C} \subseteq \mathcal{T}$ implies $\bigcup_{U \in \mathcal{C}} U \in \mathcal{T}$.[13]

Property (T2) says that the intersection of any two open sets is open, while (T3) tells us that the union of *any* collection of open sets (no matter how big the collection is) is open.[14]

Exercise 122

Show that $\mathcal{T}$ is in fact closed under all finite intersections: that is, if $U_1, \ldots, U_k \in \mathcal{T}$, then $\bigcap_{i=1}^{k} U_i \in \mathcal{T}$.

Table 5.1 A comparison of topological properties and modal
axioms and rules of inference.

Topology	Modal logic
$int(A) \subseteq A$	$\Box\varphi \to \varphi$
$int(A) \subseteq int(int(A))$	$\Box\varphi \to \Box\Box\varphi$
$int(A \cap B) = int(A) \cap int(B)$	$\Box(\varphi \wedge \psi) \leftrightarrow (\Box\varphi \wedge \Box\psi)$
$int(X) = X$	$\Box\top \leftrightarrow \top$
if $A \subseteq B$ then $int(A) \subseteq int(B)$	from $\varphi \to \psi$ deduce $\Box\varphi \to \Box\psi$

Properties (T1) and (T3) together imply that $X \in \mathcal{T}$. We also
always have $\varnothing \in \mathcal{T}$, because the empty set is the result of the
"empty union".[15]

Exercise 123

Show that $int(X) = X$ and for all $A, B \subseteq X$, $int(A \cap B) = int(A) \cap int(B)$.

Solution: page 231

Perhaps at this point you've noticed an interesting
connection.

In the left-hand column of Table 5.1 we have a list of several
of the properties we've proved of topological spaces. And in the
right-hand column, a list of axioms and a rule of inference that
look suspiciously similar to the topological expressions on the
left.[16]

The way to make this connection precise—and the corner-
stone of the topological interpretation of modality—is to
redefine the semantic interpretation of the box modality:
instead of interpreting modal formulas on frames, we can
interpret then on topological spaces;[17] and instead of defining $[\![\Box\varphi]\!]$ using a relation, we define it using the interior
operator:

$$[\![\Box\varphi]\!] = int([\![\varphi]\!]).$$

[15] That is, when $\mathcal{C} = \varnothing$, we have $\bigcup_{U \in \mathcal{C}} U = \varnothing$. It's probably best to think of this as a convention rather than a deep mathematical fact.

[16] In fact, the right-hand list (together with the axioms and rules of propositional logic) comprises an alternative sound and complete axiomatization of the class of reflexive and transitive frames.

[17] Equipped, of course, with valuations.

Before diving into the logic, however, it will be useful to look at a few concrete examples of topological spaces—if nothing else as an antidote to all the abstraction. One easy class of examples comes from taking any set X and equipping it with the full powerset as a topology: $(X, 2^X)$. Trivially, 2^X satisfies all the requirements for a topology. Any such space is called a *discrete space*. The physical intuition is that in a discrete space, every point is "far" from every other point—for each point $x \in X$, there is a *neighbourhood* of x,[18] namely $\{x\}$, that contains no other points at all. Switching to the measurement analogy, a discrete space is one in which every truth is *measurably* true—this corresponds to the fact that in such a space, for all $A \subseteq X$, $int(A) = A$.

[18] In a topological space, any open set containing x is called a *neighbourhood of x*.

At the other extreme, we can also form a topological space by starting with any set X and equipping it with the collection $\{\varnothing, X\}$ as a topology. Once again, it is easy to check that this collection satisfies all the requirements for a topology. These spaces are called *indiscrete*. The spatial picture here is one where every point is "squished right next to" every other point. Indeed, we have $int(A) = \varnothing$ whenever $A \neq X$—no point counts as robustly inside any set except for the whole space X itself. Thinking in terms of measurement, this says that nothing is measurably true except those facts that are trivially true.[19]

[19] That is, φ is only measurably true if $[\![\varphi]\!] = X$.

Moving away from these extreme examples, but still thinking simple, if $X = \{a, b\}$ then the collection $\mathcal{T} = \{\varnothing, \{a\}, X\}$ is a topology on X.[20] One easily checks that $\mathcal{T}$ satisfies the requirements for a topology. It is a useful illustration of the fact that topological structure need not be "symmetric"—in the Sierpinski space we see that a is distinguishable from b,[21] but b is *not* distinguishable from a.[22] Accordingly, we have $int(\{a\}) = \{a\}$ but $int(\{b\}) = \varnothing$.

[20] This is the smallest example of a topological space that is neither discrete nor indiscrete. It is called the Sierpinski space.

[21] In the sense that there is a measurement that can be taken at a, namely $\{a\}$, that rules out b.

[22] Because the only measurement that can be taken at b is X itself, which is also compatible with a.

One of the most important and classic examples of a topological space is the set of real numbers, $\mathbb{R}$, equipped with its "usual" topology: $\mathcal{T}_{\mathbb{R}}$, which consists of all unions of open intervals.[23] This collection is closed under arbitrary unions by construction, but is it closed under finite intersections? The answer is yes; this follows from Proposition 72, which we will see shortly, but it is a useful exercise in the basic set theory of $\mathbb{R}$ to try to prove this directly.

Exercise 124

Prove that $\mathcal{T}_{\mathbb{R}}$ is closed under finite intersections.

This way of specifying a topology—as "all unions" of sets from some other collection—is important enough to articulate in generality. Given a set X, a collection $\mathcal{B} \subseteq 2^X$ is called a **basis for a topology** if:

(B1) $\mathcal{B}$ covers X;
(B2) for all $U, V \in \mathcal{B}$ and any $x \in U \cap V$, there exists a set $W \in \mathcal{B}$ such that $x \in W \subseteq U \cap V$.

Exercise 125

Show that if $\mathcal{B}$ is closed under finite intersections, then (B2) is satisfied, but the reverse implication does not hold in general.

If $\mathcal{B}$ is a basis then $\mathcal{T}(\mathcal{B})$ is defined to be the collection of all unions of elements of $\mathcal{B}$, and this is called the *topology generated by $\mathcal{B}$*. It is indeed a topology, as we now show.

Proposition 72 *If $\mathcal{B}$ is a basis for a topology, then $\mathcal{T}(\mathcal{B})$ is a topology.*

Proof We need to show that $\mathcal{T}(\mathcal{B})$ satisfies (T1)–(T3). (T1) and (T3) are trivial, so we focus on (T2). Let $U, V \in \mathcal{T}(\mathcal{B})$; we wish to show that $U \cap V \in \mathcal{T}(\mathcal{B})$. Let $x \in U \cap V$. Since U and V are both unions of elements of $\mathcal{B}$, that means there exist $U_x, V_x \in \mathcal{B}$ such that $x \in U_x \subseteq U$ and $x \in V_x \subseteq V$. By (B2), then, there exists a set

---

[23] An open interval of $\mathbb{R}$ is any set of the form
$$(a,b) = \{x \in \mathbb{R} : a < x < b\},$$
where $a < b$. Note that the inequalities here are strict! The word "open" is suggestive here, since each open interval will indeed become an open set in $\mathcal{T}_{\mathbb{R}}$. (The notation for open intervals unfortunately clashes with that for ordered pairs; we rely on context to disambiguate.)

$W_x \in \mathcal{B}$ with $x \in W_x \subseteq U_x \cap V_x$. It follows that $W_x \subseteq U \cap V$ as well. Since we can find such a set for each $x \in U \cap V$, we can form the union

$$\bigcup_{x \in U \cap V} W_x.$$

This union clearly contains $U \cap V$, since it contains each $x \in U \cap V$; on the other hand, it is also contained in $U \cap V$, since each W_x is. Therefore

$$\bigcup_{x \in U \cap V} W_x = U \cap V.$$

This then shows that $U \cap V$ is a union of elements of $\mathcal{B}$, so by definition $U \cap V \in \mathcal{T}(\mathcal{B})$.[24]

Getting back to the real line, we can understand the collection of open intervals to be a basis for the usual topology on $\mathbb{R}$.[25] This means, for example, that the interior of any *closed* interval[26] $[a, b]$ is exactly (a, b). To see this, first note that any point $x \in (a, b)$ is in $int([a, b])$ because (a, b) itself is a neighbourhood of x contained in $[a, b]$. On the other hand, any open neighbourhood U of a must contain an open interval (c, d) that contains a,[27] and therefore U must contain points less than a (and greater than c), which implies $U \nsubseteq [a, b]$. Similarly for b.

This reasoning is worth generalizing.

Proposition 73 *Let $(X, \mathcal{T})$ be a topological space, and let $\mathcal{B}$ be a basis for $\mathcal{T}$.[28] Then $x \in int(A)$ if and only if there is some $U \in \mathcal{B}$ such that $x \in U \subseteq A$.*

In other words, in the definition of the interior, it's okay to restrict attention to *basic open sets*.[29]

Exercise 126

Prove Proposition 73.

To close this section, and in light of the previous example, we ought to say a word about *closed* sets. The easiest way to define

[24] Compare this proof to your solution to Exercise 124. How similar are they? Which one is easier?

[25] Similarly, we can view the collection of open balls as a basis for the usual topology on $\mathbb{R}^2$. (The open ball with radius ε centred at $(x, y) \in \mathbb{R}^2$ is defined to be the set

$\{(x', y') \in \mathbb{R}^2 :$
$\sqrt{(x - x')^2 + (y - y')^2} < \varepsilon\}.$

That is, the set of all points whose Euclidean distance from (x, y) is (strictly) less than ε.)

[26] A closed interval of $\mathbb{R}$ is any set of the form
$[a, b] = \{x \in \mathbb{R} : a \leq x \leq b\}$,
where $a \leq b$. Note that the inequalities here are not strict! The word "closed" is suggestive here as well, as we shall soon see.

[27] Since U is a union of open intervals.

[28] That is, $\mathcal{T}(\mathcal{B}) = \mathcal{T}$.

[29] That is, open sets that occur in some basis for the topology.

a closed set is to say it's the complement of an open set. But this is perhaps not the best for developing intuition. For that we should instead start with the *closure* operator.

Similarly to the interior operator, we can also define the closure of a set—intuitively, the closure of A consists of all those points that are "almost" in A.[30] Once again, we use the topology to capture the idea of closeness, and say that x is in the **closure** of A if, for every open set $U \in \mathcal{T}$ that contains x, we have $U \cap A \neq \varnothing$. In other words, it's impossible to "separate" x from A with an open set.[31]

When x is in the closure of A we write $x \in cl(A)$, so $cl(A)$ is the set of all points in the closure of A. It's not hard to see that x is in the closure of A iff it is not in the interior of the complement of A, and similarly, x is in the interior of A iff it is not in the closure of the complement of A.[32] This is the relationship of *duality* we have encountered before.

> ### Exercise 127
>
> Prove that $cl(A) = X \setminus int(X \setminus A)$ and $int(A) = X \setminus cl(X \setminus A)$.

It's not hard to show that the interior of any set is open[33]—in fact, the open sets are precisely the fixed points of the interior operator.[34] Thus, we call a set **closed** if it is equal to its own closure; it can then be shown that a set is closed iff its complement is open.[35]

5.2 The logic of space

At last, we get to the logic part. A **topological model** is a topological space $(X, \mathcal{T})$ together with a valuation $v \colon \mathrm{PROP} \to 2^X$. Truth in a model $M = (X, \mathcal{T}, v)$ is defined via the following semantic clauses:

$$[\![p]\!]_M = v(p)$$
$$[\![\neg\varphi]\!]_M = X \setminus [\![\varphi]\!]_M$$
$$[\![\varphi \wedge \psi]\!]_M = [\![\varphi]\!]_M \cap [\![\psi]\!]_M$$
$$[\![\Box\varphi]\!]_M = int([\![\varphi]\!]_M).$$

[30] Including points that are actually in A. Thus, the closure will be an *expanding operator*: mapping each set to a superset.

[31] In the measurement analogy: every measurement one might take at x is at least compatible with A; that is, A is *unfalsifiable*.

[32] Intuitively: x is "almost" in A iff it is not "robustly" outside of A, and x is "robustly" in A iff it is not almost outside of A.

[33] Consider the union of all open subsets of A.

[34] That is, $A \in \mathcal{T}$ iff $int(A) = A$.

[35] Compare Exercise 127.

In other words, the only difference is in the recursive clause for the box modality, as discussed. Based on the epistemic interpretation of the interior operator we've developed, we might naturally read $\Box\varphi$ as "φ is measurably true" or even "φ is knowable", though for the moment we will try to develop the mathematics as generally as possible.[36]

It's not hard to show that under these semantics, the axiom system S4 is sound with respect to the class of all topological spaces. The proof is quite similar to the other soundness proofs we have seen,[37] and is based primarily on the properties of topological spaces summarized in Table 5.1.

[36] That is, without pre-committing ourselves to a specific interpretation of $\Box$. In this spirit, a more "neutral" reading of $\Box\varphi$ might be "φ is robustly true", where the concept of "robustness" is itself left unspecified and therefore open to multiple interpretations.

[37] Namely: induction on the length of the deduction.

Exercise 128

Prove that S4 is sound with respect to the class of all topological spaces.

Exercise 129

Show that S5 is *not* sound with respect to the class of all topological spaces.

Solution: page 232

Exercise 130

A formula φ *defines* a class of topological spaces S just in case for all topological spaces $(X, \mathcal{T})$, we have

$$(X, \mathcal{T}) \vDash \varphi \text{ iff } (X, \mathcal{T}) \in S.$$

(a) What class of spaces does the formula $\varphi \to \Box\varphi$ define?
(b) Find a formula that defines the class of spaces where every closed set is open.

Solution: page 232

---

[38] This is sometimes expressed by saying that "S4 is the logic of space".

It turns out that S4 is also *complete* with respect to the class of all topological spaces.[38] To show this, we might try to construct a "canonical topological space"—and indeed this can be done and is illuminating. But we can actually prove completeness by making use of a result we already know from relational semantics, namely, that S4 is complete with respect to the class of all reflexive and transitive frames.

Here's the idea: we'll show how to transform any reflexive and transitive frame into a topological space in a way that preserves the truth values of formulas. This will then show that anything that can be refuted in a reflexive and transitive frame can be refuted in a topological space, from which the desired completeness result follows.

Proposition 74 *Let (X, R) be a reflexive and transitive frame and define $\mathcal{B}_R = \{R(x) : x \in X\}$. Then $\mathcal{B}_R$ is a basis for a topology on X.*

Proof First we need to show that $\mathcal{B}_R$ covers X. But this is easy: for each $x \in X$, by reflexivity, we have $x \in R(x)$, and by definition, $R(x) \in \mathcal{B}_R$.

Next, let $R(x), R(y) \in \mathcal{B}_R$ and take $z \in R(x) \cap R(y)$. We will show that $R(z) \subseteq R(x) \cap R(y)$, which will establish (B2). Consider any $z' \in R(z)$. We know that xRz (since $z \in R(x)$) and zRz' (since $z' \in R(z)$); thus, by transitivity, we have xRz'. This shows that $z' \in R(x)$. Since this holds for any $z' \in R(z)$, we obtain $R(z) \subseteq R(x)$. An analogous argument shows that $R(z) \subseteq R(y)$. $\square$

Let $\mathcal{T}_R = \mathcal{T}(\mathcal{B}_R)$. Thus, for each $x \in X$, the worlds accessible from x form a neighbourhood of x. In fact, they form the *smallest* neighbourhood of x, in the following sense.

Exercise 131

Show that every open neighbourhood of x in $\mathcal{T}_R$ contains $R(x)$.

Exercise 132

An *Alexandroff space* is a topological space in which arbitrary *intersections* of open sets are open.

(a) Show that in an Alexandroff space $(X, \mathcal{T})$, every point $x \in X$ has a *minimal neighbourhood*, that is, an open set $U_x \in \mathcal{T}$ such that $x \in U_x$ and $(\forall V \in \mathcal{T})(x \in V$ implies $U_x \subseteq V)$.

(b) Prove that if every point in a topological space has a minimal neighbourhood, then the space is an Alexandroff space.

(c) Conclude that $(X, \mathcal{T}_R)$ is always Alexandroff.

Solution: page 232

This way of transforming a reflexive and transitive frame into a topological space turns out to preserve the truth values of modal formulas.

Proposition 75 *Let* $M = (X, R, v)$ *be a reflexive and transitive (relational) model and let* $M' = (X, \mathcal{T}_R, v)$ *be the corresponding topological model. Then for every formula* $\varphi \in \mathcal{L}_\square$ *and each point* $x \in X$, *we have* $(M, x) \vDash \varphi$ *iff* $(M', x) \vDash \varphi$.

Proof Of course, the proof proceeds by structural induction. The base case where $\varphi \in$ PROP and the inductive steps for negation and conjunction are trivial.

So suppose inductively that the result holds for φ. We wish to show it holds for $\square\varphi$. We know that $(M, x) \vDash \square\varphi$ iff $R(x) \subseteq [\![\varphi]\!]_M$. By the inductive hypothesis, $[\![\varphi]\!]_M = [\![\varphi]\!]_{M'}$. Now $R(x) \subseteq [\![\varphi]\!]_{M'}$ clearly implies that $x \in int([\![\varphi]\!]_{M'})$. Conversely, if $x \in int([\![\varphi]\!]_{M'})$, then there is an open set $U \in \mathcal{T}_R$ such that $x \in U \subseteq [\![\varphi]\!]_{M'}$; by Exercise 131, we must have $R(x) \subseteq U$, which yields $R(x) \subseteq [\![\varphi]\!]_{M'}$. Thus we have shown that $(M, x) \vDash \square\varphi$ iff $x \in int([\![\varphi]\!]_{M'})$; since this is in turn equivalent to $(M', x) \vDash \square\varphi$, we are done. $\square$

Corollary 76 *S4 is complete with respect to the class of all topological spaces.*

Proof Suppose $\nvdash_{S4} \varphi$. Then there is some reflexive and transitive model M that refutes φ. Then by Proposition 75 the corresponding topological model M' must also refute φ. $\square$

Exercise 133

Given a topological space $(X, \mathcal{T})$, define a relation $R_\mathcal{T}$ on X by

$$x R_\mathcal{T} y \text{ iff } x \in cl(\{y\}).$$

(a) Show that $R_\mathcal{T}$ is reflexive and transitive.

(b) Must $R_\mathcal{T}$ be symmetric? Justify your answer.

(c) Let $v\colon \text{PROP} \to 2^X$. Show that if $(X, \mathcal{T})$ is Alexandroff, then for all formulas φ and each $x \in X$,

$$((X, \mathcal{T}, v), x) \vDash \varphi \text{ iff } ((X, R_\mathcal{T}, v), x) \vDash \varphi.$$

(d) Show that for every reflexive and transitive frame (X, R) we have $R_{\mathcal{T}_R} = R$, but there are topological spaces $(X, \mathcal{T})$ such that $\mathcal{T}_{R_\mathcal{T}} \neq \mathcal{T}$.

Solution: page 233

So S4 is sound and complete with respect to the class of all topological spaces.[39] Although we have already proved this result, it is fun and illuminating to try to prove completeness "directly" using a canonical model construction.[40]

[39] It turns out—though this is harder to prove—that S4 is also sound and complete with respect to much more specialized classes of spaces, like $\{\mathbb{R}\}$!

[40] Almost all of the canonical model construction for the relational case can be salvaged—the main difference is that instead of defining $\tilde{R}$, we must define $\tilde{\mathcal{T}}$, the canonical topology. For some guiding intuition, think about the Truth Lemma (which we'll still want to prove), and the fact that in a general topological space, open sets are just fixed points of the interior operator. And remember that to define a topology you only really need to define a basis for the topology!

Appendix A Mathematical Tools and Techniques

A.1 Sets, relations, and functions

In this first section of this appendix we take a quick tour through some of the most fundamental notation, terminology, and definitions in mathematics. Virtually all mathematical writing is based on these modes of expression. Accordingly, everything surveyed here can be found in countless other books, articles, and online resources, so we will keep our presentation brisk.

We begin with the notion of a *set*; intuitively, a set is just some collection of objects "considered together".[1] Since grouping things together is a fundamental way of organizing them, it's no surprise that this is a ubiquitous tool in mathematics.[2]

One way of specifying a set is to explicitly list each of the things it includes. For example, the notation

$$\{1, 2, 4, 8, 16\}$$

denotes the set that contains exactly the numbers 1, 2, 4, 8, and 16. The curly brackets tell us that we're not merely mentioning these numbers, but actively grouping them together into a single, new entity: the set that contains them. We might even name this set, for notational convenience—let's call it E, for "example".

We frequently want to talk about which objects are *in* a given set, and for this we use the symbol $\in$ as a shorthand: the expression $2 \in E$ means that the number 2 is in the set E, whereas $3 \notin E$ means that 3 is not in E.[3]

---

[1] Words like "collection" and "family" are sometimes used synonymously with "set".

[2] *Set theory* is a branch of mathematics concerned with axiomatizing the properties of sets, but for our present purposes this is overkill—we simply aim to familiarize ourselves with how sets are *used* and *manipulated* in everyday mathematical practice.

[3] We also say that 2 is (and 3 isn't) an *element of* E or a *member of* E or *contained in* E.

170

Sets are individuated by the elements they contain—that is, two sets count as the same if and only if they contain the same elements. Order doesn't matter, and duplication is irrelevant. So, for example,

$$\{1, 2, 4\} = \{4, 1, 2\} = \{1, 1, 2, 4, 4, 4\}.$$

One consequence of this is that the *empty set*—that is, the set that contains nothing at all—is unique.[4] The symbol $\varnothing$ is used to denote the empty set.

Sets don't have to be finite; for example, the set of *natural numbers*,[5]

$$\{0, 1, 2, 3, \ldots\},$$

is such a common object of study in mathematics that it has its own reserved symbol, $\mathbb{N}$.[6] We similarly have special symbols reserved for the integers $\mathbb{Z}$,[7] the rationals $\mathbb{Q}$,[8] and the real numbers $\mathbb{R}$.[9]

We can also define sets by specifying the conditions under which an object counts as being a member; for example, we might define the set $\mathbb{P}$ of all natural numbers that are *prime*:[10]

$$\mathbb{P} = \{x \in \mathbb{N} : x \text{ is prime}\}.$$

This notation is read, "the set of all x in $\mathbb{N}$ such that x is prime". Notice that we are using curly brackets once again here, even though we are not explicitly listing each individual element of the set.

There are a number of common operations on sets. One is *intersection*: starting with two sets A and B, we can produce a third set, denoted $A \cap B$, which consists of all and only those objects that are in both A and B. This is called the intersection of A and B. For instance,

$$\{1, 2, 4, 8, 16\} \cap \{4, 5, 6, 7, 8\} = \{4, 8\}.$$

When the intersection of two sets is empty, they are said to be *disjoint*.

Another common operation is *union*, denoted $A \cup B$, which consists of all and only those objects that are in either A or B:[11]

[4] You can't have two different empty sets because technically they would contain the same elements and thus count as the same set!

[5] Under some conventions, the natural numbers are defined starting from 1. But in much (perhaps most) of modern mathematics, as in this book, $0 \in \mathbb{N}$.

[6] This font is called "blackboard bold" and traditionally used for important sets of numbers.

[7] Which include the negative of every natural number, like -2 or -1000.

[8] Which include all ratios of integers, like $\frac{1}{2}$ or $\frac{-17}{13}$.

[9] Which include all infinite decimal expansions, even the ones that don't correspond to ratios of integers, like $\sqrt{2} = 1.4142\ldots$ and $\pi = 3.1415\ldots$.

[10] Recall that a number is prime if it has exactly 2 factors: 1 and itself. (Note that technically, according to this definition, 1 is not prime.)

[11] Keeping in mind, of course, that duplicates are ignored.

$$\{1,2,4,8,16\} \cup \{4,5,6,7,8\} = \{1,2,4,5,6,7,8,16\}.$$

The *complement* of a set, roughly speaking, consists of everything that's *not* in the set. This is always relative to some (often implicit) background set; for instance, it may seem obvious that the complement of $\mathbb{P}$ includes, say, the number 4, but what about the number π? This issue here boils down to whether we are taking the complement of $\mathbb{P}$ relative to the set of natural numbers or some other set (like $\mathbb{R}$). It's helpful to be explicit about this; thus we define $A \setminus B$ to consist of those elements of A that are not in B.[12] For example,

[12] This operation is sometimes called *set minus*.

$$\{1,2,4,8,16\} \setminus \{4,5,6,7,8\} = \{1,2,16\}.$$

In this notation, the complement of $\mathbb{P}$ relative to the background set of natural numbers is denoted $\mathbb{N} \setminus \mathbb{P}$, whereas its complement relative to the reals is written $\mathbb{R} \setminus \mathbb{P}$. Of course, we then have $\pi \in \mathbb{R} \setminus \mathbb{P}$ but $\pi \notin \mathbb{N} \setminus \mathbb{P}$. When a background set X is clear from context, the complement of A in X, namely $X \setminus A$, is sometimes written A^c.

Summarizing, we have:

$$A \cap B = \{x : x \in A \text{ and } x \in B\}$$
$$A \cup B = \{x : x \in A \text{ or } x \in B\}$$
$$A \setminus B = \{x : x \in A \text{ but } x \notin B\}.$$

Exercise 134

Prove the following. Draw some pictures if you get stuck!

(a) $A \cup (B \cap C) = (A \cup B) \cap (A \cup C)$
(b) $A \setminus B = A \setminus (A \cap B)$
(c) $A \cap A^c = \varnothing$
(d) $(A \cup B)^c = A^c \cap B^c$

---

[13] Two trivial instances of this definition are worth noting: for any set A, we have both $A \subseteq A$ and $\varnothing \subseteq A$. In the latter case the containment is said to be *vacuous*.

We say that A is a *subset* of B and write $A \subseteq B$ when every element of A is also in B.[13] For example,

$$\{2,4,8\} \subseteq \{1,2,4,8,16\},$$

but

$$\{2,4,8\} \not\subseteq \mathbb{P}.$$

When $A \subseteq B$ we also sometimes say that B is a *superset* of A, and write $B \supseteq A$. Note that $A \subseteq B$ and $B \subseteq A$ if and only if $A = B$. Indeed, establishing mutual containment is often a good way to prove that two sets are equal.

In many contexts it is useful to collect together *all* the subsets of a given set A; this is called the *powerset* of A and is denoted 2^A.[14] For instance,

$$2^{\{2,4,8\}} = \{\varnothing, \{2\}, \{4\}, \{8\}, \{2,4\}, \{2,8\}, \{4,8\}, \{2,4,8\}\}.$$

[14] This may seem like odd notation, and indeed many authors use a notation more like $\mathcal{P}(A)$ to denote the powerset. However, there is some rationale behind this way of writing it; see Exercise 138(b).

Exercise 135

Prove the following. Draw more pictures!

(a) If $A \subseteq B$ and $B \subseteq C$ then $A \subseteq C$.
(b) $A \cap B \subseteq A$ and $A \subseteq A \cup B$.
(c) If $A \subseteq B$ then $B^c \subseteq A^c$.

Exercise 136

It may be tempting to think that we always have $A \subseteq 2^A$. However, this is not so; explain why.

Solution: page 234

Although elements of sets are not understood to be in any particular order, in many contexts we do want to encode order. Thus we introduce the notion of an *ordered pair*, denoted (x,y), which intuitively is just like the set $\{x,y\}$ except the order matters. In particular, this means that even though $\{2,7\} = \{7,2\}$, we have $(2,7) \neq (7,2)$. Naturally, this idea of imbuing a set with an order can be extended from pairs to ordered triples (x,y,z), ordered quadruples (w,x,y,z), and so on; in general these are called *n*-tuples, and always denoted using the same round-bracket notation: $(x_1, x_2, \ldots, x_n)$.

Exercise 137

Even though sets don't inherently encode order, there *is* a way to use sets to represent order, and thus for instance capture the difference between $(2, 7)$ and $(7, 2)$. Can you invent it?

Order allows us to define another crucial operation on sets, the *Cartesian product*, denoted $A \times B$.[15] The elements of $A \times B$ are precisely the ordered pairs of elements from A and B, in that order:

$$A \times B = \{(a, b) \ : \ a \in A \text{ and } b \in B\}.$$

For example,

$$\{1, 2\} \times \{1, 2, 3\} = \{(1, 1), (1, 2), (1, 3), (2, 1), (2, 2), (2, 3)\}.$$

We similarly define the Cartesian product of n sets:

$$A_1 \times A_2 \times \cdots \times A_n = \{(a_1, a_2, \dots, a_n) \ : \ a_i \in A_i \text{ for each } 1 \leq i \leq n\}.$$

When taking the Cartesian product of a set with itself, we often use "exponent" notation, so $A \times A$ is instead written A^2, and similarly for A^n.[16]

Finally, for finite sets it is often useful to refer to how many elements a given set A contains; this is called the *cardinality* of A and denoted $|A|$.[17] So for example we have

$$|\{2, 4, 8\}| = 3,$$

and

$$|2^{\{2, 4, 8\}}| = 8.$$

---

[15] As with the powerset notation, there is a reason the symbol used to denote the Cartesian product is also one commonly used for multiplication; see Exercise 138(c).

[16] Thus, for example, $\mathbb{R}^2$ is the set of all ordered pairs of real numbers, which can be pictured as an infinite plane; similarly $\mathbb{R}^3$ is the set of all ordered triples of real numbers, adding a third axis to the visualization above to give us three-dimensional space.

[17] The idea of cardinality can also be applied to infinite sets, though in this context there are some interesting subtleties to address. See Appendix A.4.

Exercise 138

Assuming A and B are finite sets, prove the following.

(a) $|A \cup B| = |A| + |B| - |A \cap B|$

(b) $|2^A| = 2^{|A|}$

(c) $|A \times B| = |A| \cdot |B|$

(d) $|A^n| = |A|^n$

Mathematics is not just about grouping objects together;[18] we also wish to reason about relationships between objects. For example, the notation $2 < 3$ expresses a familiar relationship between the numbers 2 and 3, namely, that the former is less than the latter. Besides being part of the bedrock of mathematical notation and formalism, relations are of central importance in the study of logic, and especially modal logic.[19]

As it happens, sets can be (and are!) leveraged to formalize even this notion:[20] relationships between objects can be encoded using ordered pairs, or more precisely, *sets* of ordered pairs. The basic idea is to identify a relation with the set of all pairs of objects that actually stand in that relationship to one another. For example, the relation "$<$" on the natural numbers can be identified with the set

$$\{(a, b) \in \mathbb{N}^2 \ : \ a \text{ is less than } b\}.$$

On this view, the symbol "$<$" literally denotes a set.[21] Naturally, there can also be relationships that apply to more than two objects at a time.[22] Thus in general we define an *n-ary relation R* to be any subset of an *n*-fold Cartesian product:[23]

$$R \subseteq A_1 \times \cdots \times A_n.$$

When $R \subseteq X^2$, R is said to be a binary relation *on X*; such relations are ubiquitous in mathematics, so much so that many potential properties of such relations have names, and it is common to write xRy as an abbreviation for $(x, y) \in R$.[24] We will use both notations interchangeably.

For reference, we provide the names and definitions of some common properties of relations.

- *Reflexive:* for all $x \in X$, xRx.
- *Symmetric:* for all $x, y \in X$, if xRy then yRx.
- *Transitive:* for all $x, y, z \in X$, if xRy and yRz then xRz.

[18] At least not in ordinary parlance.

[19] Indeed, they're a fundamental part of possible world semantics.

[20] This begins to give a sense as to why set theory is often viewed as the foundation of mathematics: seemingly everything can be formalized with sets. See also Exercise 137 and sidenote 32.

[21] And so, if the reader will excuse the somewhat perverse notation, an expression like $2 < 3$ is really shorthand for

$$(2, 3) \in \ <.$$

[22] Like "x is between y and z".

[23] The 2-ary relations are typically called *binary* relations, and 3-ary relations *ternary* relations.

[24] This is called *infix notation* and it is a very common convention; for example, we all use it for the relations $<$ and $=$.

25 The "or" here is not meant to be exclusive, that is, it permits both to be true.

- *Complete:* for all $x, y \in X$, either xRy or yRx.[25]
- *Serial:* for all $x \in X$, there is some $y \in X$ such that xRy.

And here are some less common properties.

- *Irreflexive:* for all $x \in X$, it is not the case that xRx.
- *Asymmetric:* for all $x, y \in X$, if xRy then it is not the case that yRx.
- *Negatively transitive:* for all $x, y, z \in X$, if neither xRy nor yRz, then it is not the case that xRz.
- *Euclidean:* for all $x, y, z \in X$, if xRy and xRz, then yRz.
- *Acyclic:* for all n, there are no n-cycles, where an *cycle* is a collection of elements $x_1, \ldots, x_n \in X$ such that $x_1 R x_2$, $x_2 R x_3$, $\ldots$, $x_{n-1} R x_n$, and $x_n R x_1$.

Exercise 139

Consider the following sets and relations.

(A) $X_A = \{\text{words in the dictionary}\}$;
   $w R_A w'$ iff the last letter of w matches the first letter of w'.
(B) $X_B = \mathbb{R}$;
   $x R_B y$ iff $|2x - y| < 1$.
(C) $X_C = \mathbb{N} = \{0, 1, 2, \ldots\}$;
   $x R_C y$ iff $3x + y$ is even.

For each of the following, justify your answers.

(a) Is R_A transitive? Reflexive? Acyclic?
(b) Is R_B reflexive? Irreflexive? Symmetric?
(c) Is R_C symmetric? Transitive? Negatively transitive?

Solution: page 234

Exercise 140

Give examples of relations with the following properties:

(a) reflexive but not transitive
(b) transitive but not reflexive

> (c) neither symmetric nor transitive nor reflexive
> (d) symmetric and transitive but not reflexive.

Various implications hold between different properties of relations. For example, we have the following.

Proposition 77 *If R is both transitive and irreflexive, it is also asymmetric.*

Proof Suppose that R is transitive and irreflexive, and let $x, y \in X$ be such that $(x, y) \in R$.[26] We wish to show that $(y, x) \notin R$. So suppose, by way of contradiction, that yRx. Then by transitivity we can deduce that xRx, which violates irreflexivity, a contradiction. □

[26] If there are no such elements in X then R is trivially asymmetric.

It's even easier to prove (though perhaps a bit trickier to notice) that asymmetry implies irreflexivity.

Proposition 78 *If R is asymmetric, it is also irreflexive.*

Proof Suppose that R is asymmetric, and let $x \in X$. Then we cannot have $(x, x) \in R$, because if we did, by asymmetry we could also deduce that $(x, x) \notin R$, a contradiction. Thus we conclude that it is not the case that xRx. □

In Proposition 77 we can also demonstrate that neither assumption can be dropped.

Proposition 79 *Neither transitivity nor irreflexivity alone implies asymmetry.*

Proof For the first claim, it suffices to show that there is a relation that is both transitive and not asymmetric. For this, the relation of equality on (say) the natural numbers $\mathbb{N}$ is sufficient, since it is transitive but not asymmetric.

For the second claim, again it suffices to show that there is a relation that is both irreflexive and not asymmetric. For this, consider the set $X = \{1, 2\}$ and the relation $R = \{(1, 2), (2, 1)\}$, and observe that it is clearly irreflexive yet not asymmetric. □

There are a number of other implications and non-implications that hold between various combinations of these properties, and exploring them is excellent practice.

Exercise 141

Prove the following.

(a) Acyclic implies asymmetric.
(b) Complete implies reflexive.
(c) Serial, symmetric, and transitive implies reflexive, but no two of these assumptions alone is enough to imply reflexivity.
(d) Reflexive and Euclidean implies symmetric, but neither suffices alone.
(e) Asymmetric and negatively transitive implies transitive.
(f) Asymmetric and transitive does not imply negatively transitive.

Exercise 142

Say that two properties of relations are *independent* if neither implies the other. Show the following.

(a) Asymmetry and negative transitivity are independent.
(b) Transitivity and negative transitivity are independent.
(c) Independence is not transitive.

Solution: page 235

Besides grouping objects together and reasoning about their relationships, mathematicians also like to *do things* to objects, namely, transform them into new objects in various ways. Such transformations are called *functions*; aside from numbers, functions are perhaps the most familiar kind of mathematical object. Like relations, they are part of the mathematical bedrock and come up frequently in this book, in many different contexts. We therefore close this section with a brief look at what functions are and how they fit into the formal set-theoretic framework introduced above.

The basic idea of a function is that is tells us how to transform certain types of inputs into certain types of outputs.[27] Naturally, the "allowed" inputs and outputs are presented as sets: we write

$$f: X \to Y$$

to indicate that f is a function that takes inputs from the set X and, for each such input, produces an output in the set Y. We call X and Y the *domain* and the *codomain* of f, respectively.[28] Given any $x \in X$, we write $f(x)$ to denote the output that f produces when fed x as input.[29] We then read "$f(x) = y$" as "f maps x to y" or "the value of f at x is y".[30]

For example, the function $g: \mathbb{R} \to \mathbb{R}$ defined by

$$g(x) = x^2 + 1$$

takes as input any real number and produces a real number as output, namely, the square of the input plus one.[31] Thus, the domain and codomain of g are both $\mathbb{R}$, while its range is $\{y \in \mathbb{R} : y \geq 1\}$, since $x^2 \geq 0$ no matter what x is.

Functions can be (and are) viewed as a special kinds of binary relations. Specifically, a function $f: X \to Y$ can be identified with the relation

$$R_f = \{(x,y) \in X \times Y : f(x) = y\},$$

which is just the set of all input–output pairs determined by applying f to every element in its domain.[32] This view of a function treats it essentially as a record of every transformation it might ever perform.[33] Conversely, any relation $R \subseteq X \times Y$ can be understood as a function from X to Y provided each $x \in X$ is related via R to *exactly one* $y \in Y$. In this case, we can identify R with the function $f_R: X \to Y$ defined by:

$$f_R(x) = y \text{ iff } (x,y) \in R.$$

[27] For this reason, functions are sometimes called *maps* or *mappings*, since intuitively they tell you how to get from some starting point (an element of X) to some destination (an element of Y).

[28] Sometimes the word "range" is used instead of "codomain", but this can be confusing because the *range* of f is also used to refer to the subset of Y consisting of those elements that actually occur as outputs of f. In this book we adopt this latter definition.

[29] Which of course means that $f(x) \in Y$.

[30] This is also sometimes written $x \overset{f}{\mapsto} y$.

[31] So, for example, we can write (true) expressions like $g(2) = 5$, $g(0) = 1$, and $g(-0.5) = 1.25$, and say (true) things like "g maps 2 to 5" or "the value of g at 0 is 1", etc.

[32] Once again vindicating the "everything is a set" perspective on mathematics.

[33] This is sometimes called an *extensional* view; it contrasts with an *intensional* view in which, roughly speaking, functions are identified with rules or processes. To see the difference, consider the rules "multiply by two then add six" versus "add three then multiple by two". Although these are two different rules in some sense, they produce exactly the same input–output pairs. Are the functions determined by these rules the same, or different? Modern mathematics, which tends to favour extensional approaches, says they are the same function, but it is possible to explore other notions of "sameness".

Notice that the uniqueness condition on R ensures that f_R produces a unique output for each input $x \in X$, which was (implicitly) part of the original notion of what a function is.

[34] Or sometimes *onto*.

A function $f\colon X \to Y$ is called *surjective*[34] if its range covers the entire codomain, or in other words, if every $y \in Y$ is mapped to by some $x \in X$:

$$(\forall y \in Y)(\exists x \in X)(f(x) = y).$$

[35] Or sometimes *one-to-one*.

It is called *injective*[35] if no two distinct elements of X get mapped to the same output in Y:

$$(\forall x, x' \in X)(x \neq x' \text{ implies } f(x) \neq f(x')).$$

Functions that are both injective and surjective are called *bijective*. Bijections play a very important role in mathematics— intuitively, they establish a "rigid alignment" between the domain and codomain by pairing off elements of the former with elements of the latter with no overlap (injectivity) and nothing left out (surjectivity). This turns out to be the crucial tool used to define notions of cardinality for infinite sets.[36]

[36] See Appendix A.4.

If $f\colon X \to Y$ and $g\colon Y \to Z$ are functions, they can be "chained together" to form a new function that takes inputs in X and produces outputs in Z by first applying f and then applying g to the output of f. That is, $x \in X$ is first mapped to $f(x) \in Y$, which is then taken to $g(f(x)) \in Z$. This new function is called the *composition* of g and f and written $g \circ f$:

$$(g \circ f)(x) = g(f(x)).$$

Exercise 143

Show that the composition of two injective functions is injective and the composition of two surjective functions is surjective.

It is also frequently useful to consider functions "in reverse": mapping outputs back to inputs. These are called *inverses*. A wrinkle here is that since a function $f: X \to Y$ may map different inputs to same output, its inverse may not actually be a function from Y to X, since it may associate a single element of Y with multiple elements of X. We thus define

$$f^{-1}(y) = \{x \in X \ : \ f(x) = y\};$$

in other words, the inverse "function" f^{-1} associates each element $y \in Y$ with a *set* of elements in X, namely, the set of all $x \in X$ that f maps to y.[37]

If $f: X \to Y$ is injective and $f(x) = y$, then by definition x is the only element in the domain that maps to y, so $f^{-1}(y) = \{x\}$. In this case we often write $f^{-1}(y) = x$ to emphasize that f^{-1} really does behave like a function from Y to X.[38]

One final bit of notation and terminology: given $f: X \to Y$ and $A \subseteq X$, it is common to abuse notation[39] slightly and write $f(A)$ to denote every element mapped to by anything in A, which is sometimes called the *image* of A:

$$f(A) = \{y \in Y \ : \ (\exists x \in A)(f(x) = y)\}.$$

Similarly, given $B \subseteq Y$, we write $f^{-1}(B)$ to denote every element that maps to anything in B, called the *preimage* of B:

$$f^{-1}(B) = \{x \in X \ : \ f(x) \in B\}.$$

Exercise 144

Let $f: X \to Y$. Show that for all $x \in X$ we have $|f^{-1}(f(x))| \geq 1$, and for all $y \in Y$ we have $|f(f^{-1}(y))| \leq 1$.

Exercise 145

Show that if $f: X \to Y$ is a bijection, then f^{-1} can be viewed as a function from Y to X, and it is also a bijection.

A.2 Equivalence relations and partitions

An *equivalence relation* is any relation that is reflexive, symmetric, and transitive. Intuitively, an equivalence relation captures

[37] For example, for the function g defined above by $g(x) = x^2 + 1$, we have $g^{-1}(5) = \{2, -2\}$, since $g(2) = 5$ and also $g(-2) = 5$. We also have $g^{-1}(1) = \{0\}$ and $g^{-1}(-2) = \varnothing$.

[38] Note, though, that unless f is surjective, f^{-1} will be undefined on some elements of Y. So more properly f^{-1} should be viewed as a function from the range of f to X.

[39] "Abusing notation" is exceedingly common in mathematics. In this case the "abuse" is that f is only supposed to take inputs that are individual elements of X, not sets of elements. That being said, notation is supposed to work *for us*, not the other way around—if there is some way we can make our lives easier and our formulas more concise and readable by systematically bending the rules, why shouldn't we take it?

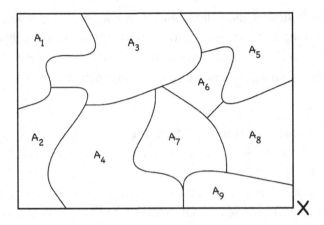

Figure A.1 A sample partition $\Pi = \{A_1,\ldots,A_9\}$ of X.

a notion of *sameness*. The ultimate notion of sameness is, of course, identity, as captured by the relation of equality (=) itself.[40] However, in many contexts we want to treat two things that are not literally equal as being "the same for present purposes". For instance, when we say that two people have the "same" birthday, we don't typically mean that they were born on the exact same day as one another; rather, we count two people as having the same birthday if they were born on the same *calendar* day. In other words, we essentially divide the continuity of time into 366 chunks,[41] and treat two people as having the same birthday provided their moments of birth fall into the same chunk.

This idea of dividing time into "chunks" can be generalized to any set X. A *partition* of X is a collection of nonempty subsets $\Pi = \{A_i\}$, such that

(1) $\bigcup_i A_i = X$;[42]
(2) if $A_i \neq A_j$ then $A_i \cap A_j = \varnothing$.[43]

Elements $A_i \in \Pi$ are called *cells* of the partition. So Π is a collection of mutually disjoint subsets of X that covers X—like the pieces of a jigsaw puzzle.

[40] Note that equality is also reflexive, symmetric, and transitive.

[41] So, for example, June 4th, 1985, at 2:33 p.m. falls into the same "chunk" as June 4th, 1966, at 7:27 a.m.

[42] Π *covers* X.

[43] The elements of Π are *mutually disjoint*.

Given a partition Π of X, we can define a binary relation E_Π on X capturing "sameness" or "equivalence" with respect to the partition just as we did in the birthday case:

$$x E_\Pi y \Leftrightarrow \text{for some } A_i \in \Pi, x, y \in A_i.$$

Proposition 80 *The relation E_Π is an equivalence relation.*

Proof We must show that E is reflexive, symmetric, and transitive. For reflexivity, we have $x E_\Pi x$ just in case for some $A_i \in \Pi$, $x \in A_i$, which follows from the fact that Π covers X. Symmetry is immediate from the definition of E_Π. For transitivity, suppose that $x E_\Pi y$ and $y E_\Pi z$; we must show that $x E_\Pi z$. By definition, we know that there is some $A_i \in \Pi$ such that $x, y \in A_i$, and some $A_j \in \Pi$ such that $y, z \in A_j$. So we have $y \in A_i \cap A_j$, from which it follows (by disjointness) that $A_i = A_j$; hence $x, z \in A_i$, which shows that $x E_\Pi z$, as desired. □

So every partition gives rise to an equivalence relation as above. Perhaps the more surprising fact is that *every* equivalence relation arises in this way; in fact, the correspondence $\Pi \mapsto E_\Pi$ is a bijection between the class of partitions and the class of equivalence relations.[44] To see this, we'll show how, starting with an equivalence relation, we can construct a partition that gives rise to it.

How does an equivalence relation E on X give rise to a partition? The intuition is simple: group together the elements of X into "bundles" based on which ones are equivalent. More precisely, for each $x \in X$, define the *equivalence class* of x to be the set

$$[x]_E = \{y \in X : y E x\}.$$

Thus, the equivalence class of x is just the set of all the elements of X that are E-related to x. Now define $\Pi_E = \{[x]_E : x \in X\}$.[45]

Proposition 81 *The set Π_E is a partition of X.*

Proof This follows immediately from Exercise 146. □

[44] That is, there is a one-to-one correspondence between partitions and equivalence relations. Intuitively, partitions and equivalence relations are the same thing, just viewed from different perspectives.

[45] Note that distinct elements x and y may be such that $[x]_E = [y]_E$, but since Π_E is a set, it doesn't matter how many times we list the same redundant element, it gets counted only once.

Exercise 146

Prove the following.

(a) For each $x \in X$, $x \in [x]_E$.

(b) xEy if and only if $[x]_E = [y]_E$.

(c) $[x]_E \neq [y]_E$ implies $[x]_E \cap [y]_E = \varnothing$.

Solution: page 236

Exercise 147

Show that $\Pi_{E_\Pi} = \Pi$ and $E_{\Pi_E} = E$.

Since elements that belong to the same equivalence class are related by E, we can think of the set Π_E as what we get if we start with X and "forget" the distinction between E-related elements.[46] And this is exactly what the original motivation was—to treat elements that are "the same for present purposes" (i.e., equivalent according to E) *as if* they were actually equal.

[46] Π_E is sometimes called the *quotient* of X by E, written X/E.

A.3 Induction

The basic goal of mathematical induction is to prove that some property of interest holds true for *every* natural number; that is, to prove that it's true for 0, for 1, for 2, and so on forever, for every natural number no matter how large.

Of course, we cannot prove an infinite number of things like this by writing down our reasoning for each case separately—we would never finish. To prove that something is true for infinitely many cases we need somehow to take advantage of the underlying structure of the cases, so that we can essentially apply the same argument over and over again to all of them.

Sometimes doing so is fairly straightforward. For example, consider the following claim: "Every even number, when squared, is divisible by 4." We can prove this pretty directly by first considering what an "even number" is: it is simply any natural number that can be written in the form $2n$ for some natural number n. Thus, if we are given an arbitrary even number, we know it must look like $2n$ for some n, and so we know its square is equal to $(2n)^2 = 4n^2$. This shows that its square is four times some number (namely n^2), which is exactly what "divisible by 4" means.

Induction provides a different approach to proving properties of natural numbers, one that turns out to be quite powerful. The high-level idea is the following: suppose we want to show that some property P is true of every natural number; we'll write $P(n)$ to mean that the property holds for the number $n \in \mathbb{N}$. There are only two steps. We begin by proving $P(0)$—that is, that the property holds for the first natural number, 0. This is called the *base case*. Second, for the *inductive step*, we prove the following: for each natural number n, *if* $P(n)$ is true, so is $P(n + 1)$. In other words, the truth of P is "preserved" by the operation of "adding 1".

The two steps above create a sort of chain reaction. We know $P(0)$ is true (from the base case), and from this it follows that $P(1)$ must be true too (by the inductive step). But then since $P(1)$ is true, $P(2)$ must be true as well (again using the inductive step), and similarly we deduce that $P(3)$ is true, $P(4)$ is true, etc.

The hardest part of doing induction is usually the inductive step, not only because it typically requires the most work, but because conceptually it is easy to get confused about it. To establish the inductive step we must *assume* $P(n)$ is true and then use this to prove that $P(n + 1)$ is true. But assuming that $P(n)$ is true for some arbitrary n feels like cheating! It feels a lot like assuming the very thing we're trying to prove, namely, that $P(n)$ is true for all n. The key difference is that for the inductive step we're only temporarily assuming $P(n)$ in order to prove that *if* it *were* true, it would force $P(n + 1)$ also to be true. This gives us the chain reaction.[47]

Let's see how this works with a simple example.

Proposition 82 *For all natural numbers $n \geq 1$, the sum of the first n odd numbers is equal to n^2.*

[47] If you feel confident that you understand this, Exercise 155 is a good way to test yourself!

[48] So $S(2)$ is the claim that
$1 + 3 = 2^2$, $S(5)$ is the claim
that $1 + 3 + 5 + 7 + 9 = 5^2$, etc.

Proof Let $S(n)$ denote the claim that the sum of the first n odd numbers, $1 + 3 + 5 + \cdots + (2n - 1)$, is equal to n^2.[48] Our task is to prove the (perhaps somewhat surprising) fact that $S(n)$ is true for every n—that is, no matter what natural number n you pick, if you add up the first n odd numbers, the result is n^2.

It's not immediately obvious how to do this, but induction provides a kind of recipe that we can try to use to organize our approach. The base case (as often happens) is trivial: $S(0)$ is the statement that $0 = 0^2$, that is, the sum of the first 0 odd numbers is 0. Great.

For the inductive step, we must show that for each n, $P(n)$ implies $P(n + 1)$. That is, assuming

$$1 + 3 + 5 + \cdots + (2n - 1) = n^2, \tag{A.1}$$

we want to prove that

$$1 + 3 + 5 + \cdots + (2n - 1) + (2n + 1) = (n + 1)^2. \tag{A.2}$$

Since we are taking equation (A.1) for granted, we also know that

$$1 + 3 + 5 + \cdots + (2n - 1) + (2n + 1) = n^2 + (2n + 1), \tag{A.3}$$

since this equality follows by adding $(2n + 1)$ to both sides. But notice that simple multiplication shows that

$$(n + 1)^2 = n^2 + 2n + 1,$$

and substituting this into equation (A.3) yields equation (A.2), the desired result.   □

As is the case for essentially all of mathematics, the best way to get comfortable with induction is by doing it over and over again. We consider two more examples before turning the task over to the reader with a number of exercises.

Proposition 83 *Prove by induction the following formula for the sum of powers of 2:*[49]

$$\sum_{i=0}^{n} 2^i = 2^{n+1} - 1.$$

Proof When $n = 0$, we have $\sum_{i=0}^{0} 2^i = 2^0 = 1$ and $2^{0+1} - 1 = 2 - 1 = 1$, so equality holds in the base case. Now suppose inductively the result holds for n. We want to show it also holds for $n + 1$. Observe that

$$\sum_{i=0}^{n+1} 2^i = 2^{n+1} + \sum_{i=0}^{n} 2^i = 2^{n+1} + 2^{n+1} - 1 = 2 \cdot 2^{n+1} - 1 = 2^{n+2} - 1,$$

which establishes the desired equality. □

Proposition 84 *If R is a binary relation on X that is both transitive and irreflexive, then it is also acyclic.*[50]

Proof Suppose that R is transitive and irreflexive. We will prove by induction that for every $n \geq 1$, there is no n-cycle. First the base case, $n = 1$.[51] In this case an n-cycle is just a point x_1 with $x_1 R x_1$; this is impossible by irreflexivity. So we have established that there are no 1-cycles.

Now suppose inductively that there are no n-cycles; we must show that there are no $(n + 1)$-cycles. Suppose for contradiction that $x_1, \ldots, x_{n+1}$ is an $(n + 1)$-cycle. Then we know (among other things) that $x_n R x_{n+1}$, and $x_{n+1} R x_1$. By transitivity, it follows that $x_n R x_1$; thus, $x_1, \ldots, x_n$ is an n-cycle.[52] Since this contradicts the inductive hypothesis, we can conclude that there are no $(n + 1)$-cycles, as desired. This completes the induction. □

[50] See Appendix A.1 for the definitions of these properties of relations.

[51] We start with 1 and not 0 because we did not define 0-cycles.

[52] It's a good idea to draw a picture of this!

Exercise 148

Prove by induction the following formula for the sum of an arithmetic series:

$$\sum_{i=0}^{n} (a + id) = \frac{n+1}{2} (2a + nd).$$

Exercise 149

Prove by induction the following formula for the sum of the first n squares:

$$\sum_{i=1}^{n} i^2 = \frac{n(n+1)(2n+1)}{6}.$$

Exercise 150

Factorial notation is defined recursively for natural numbers greater than 0 as follows: $1! = 1$, and $(n + 1)! = (n + 1) \cdot n!$. Prove that for all $n \geq 5$ we have $n! > 2^n > n^2$.

Solution: page 236

Exercise 151

Prove that for all natural numbers n,

$$\sum_{i=0}^{n} i^3 = \left(\sum_{i=0}^{n} i \right)^2 .$$

Exercise 152

Prove that every multiple of 9 has the property that, if you add its digits, the resulting sum is also a multiple of 9. (For example, $9 \cdot 142 = 1278$, and $1 + 2 + 7 + 8 = 18 = 9 \cdot 2$.)

Exercise 153

Prove that every number of the form $2^{2n} - 1$ (for $n \in \mathbb{N}$) is divisible by 3.

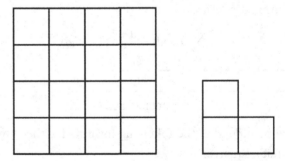

Figure A.2 A 4×4 surface and an L-tile.

Exercise 154

Figure A.2 shows, on the left, a surface made up of equal-sized squares arranged in a 4×4 grid, and on the right, an "L-tile" composed of three such squares in the L-shape depicted. An *L-tiling* of a surface made up of (same-sized) squares is a way of arranging L-tiles on top of it such that (1) each square of each L-tile is directly over some square of the underlying surface, and (2) every square of the underlying surface is covered by exactly one L-tile.

(a) Consider the surface obtained by removing one corner square from the 4×4 surface. Prove that there is an L-tiling of this surface.
(b) Prove that there is *no* L-tiling of the 4×4 surface, or of any $2^n \times 2^n$ surface for $n \geq 1$.
(c) Prove that for every $n \geq 1$, there is an L-tiling of the surface obtained by removing 1 corner square from the $2^n \times 2^n$ surface.

Hint: page 237

Exercise 155

Consider the claim: "All horses are the same colour." This is equivalent to the following universally quantified claim: "For every natural number $n \geq 1$, if H is a herd of exactly n horses, then every horse in H is the same colour." Let's try to prove this claim by induction.

First the base case, $n = 1$. Obviously any herd H consisting of only one horse is such that all horses in H are the same colour.

Now suppose inductively that the result is true for n; we want to prove it for $n + 1$. Let H be a herd consisting of exactly $n + 1$ horses. Take one horse out of H, call it a, and let H' denote the remaining herd.

> Clearly H' is a herd of size n, so by the inductive hypothesis, all the horses in H' are the same colour. Of course, this still leaves horse a unaccounted for. However, we can take some other horse b out of the original herd H (leaving a inside), and call the resulting herd H''. Again, H'' has exactly n horses in it, so once again, by the inductive hypothesis, all the horses in H'', including a, are the same colour. Therefore all the horses in the full herd H must be the same colour. This completes the induction.
>
> Obviously something has gone wrong. What?
>
> Hint: page 237

One final note: it is sometimes convenient in the course of writing inductive proofs to make a stronger inductive hypothesis: instead of assuming just $P(n)$ is true (and trying to establish $P(n + 1)$), we assume all of $P(0), P(1), \ldots, P(n)$ are true. This is sometimes called *strong induction*.[53]

[53] Though as we are about to observe, the technique is not actually any stronger than regular induction, so this is a bit of a misnomer.

What justifies this? After all, we cannot typically help ourselves to stronger assumptions just because they're convenient. In this case, however, it is induction itself that justifies the manoeuver.

Theorem 85 *If "for all n, $P(n)$" can be proved by strong induction, then it can also be proved by regular induction.*

Proof Let $Q(n)$ stand for the conjunction of all the statements $P(0), P(1), \ldots, P(n)$. Then it is easy to see that if $P(n)$ is true for all n, so is $Q(n)$, and vice versa. So if we can prove "for all n, $Q(n)$" by regular induction, we will be done.

As it happens, proving "for all n, $Q(n)$" by regular induction is equivalent to proving "for all n, $P(n)$" by strong induction. To see this, first observe that the base cases are identical, since $Q(0)$ is the same as $P(0)$. Now the inductive step for $Q(n)$ (as usual) has us assume $Q(n)$ and try to derive $Q(n+1)$; in other words, we assume $P(0), P(1), \ldots, P(n)$ and must derive $P(0), P(1), \ldots, P(n), P(n + 1)$. Since the only novel part of what we are trying to derive in this case is $P(n + 1)$, we can see that this is equivalent to deriving

$P(n + 1)$ from $P(0), P(1), \ldots, P(n)$, which is exactly the inductive step in a *strong* induction for P.   □

A.4 Infinity

Infinity is weird.[54] Entire books have been written on the subject. For the most part in this book we don't have to think too hard about infinity, but here and there we insist or point out that certain infinite sets are "countable", and it's useful to have some idea of what that means and why it's important.

In a nutshell: countable sets are those whose elements can be *listed*. More precisely, X is **countable** if there exists a (finite or infinite) sequence $x_0, x_1, x_2, \ldots$ such that each $x_i \in X$, and for every $x \in X$ there is some i such that $x_i = x$.[55] Clearly, any finite set has this property. But, perhaps surprisingly, not all infinite sets do.

The natural numbers $\mathbb{N}$ are an easy example of a countably infinite set: indeed, the elements of $\mathbb{N}$ are essentially already presented in list form: let $x_0 = 0$, $x_1 = 1$, and in general $x_i = i$, and of course the (infinite) sequence $x_0, x_1, \ldots$ is a list of all and only the elements of $\mathbb{N}$. This is no accident; indeed, the very concept of a "list" basically already has $\mathbb{N}$ baked in: the indices of a sequence are just the natural numbers themselves. In this sense, $\mathbb{N}$ can be thought of as the "original" or "canonical" countably infinite set.

This is the reason why countably infinite sets are so important: they are exactly the infinite sets that are "similar enough to $\mathbb{N}$" to be amenable to inductive proofs and recursive definitions. There are many examples of this scattered throughout the pages of this book.

Lots of infinite sets besides $\mathbb{N}$ are countable. For example, the integers $\mathbb{Z}$ are countable. This may be somewhat counterintuitive, since $\mathbb{Z}$ is a strict superset of $\mathbb{N}$, and if we start listing elements of $\mathbb{Z}$ as we did with $\mathbb{N}$, namely

$$0, 1, 2, 3, 4, 5, \ldots,$$

[54] If you don't believe me, look up the Ross–Littlewood paradox.

[55] Whether we start numbering this sequence with 0 or 1 makes no real difference, so we'll freely do either one as elegance demands.

we will never list all the integers since we'll miss the negative ones. But we *can* list all integers if we simply present them in a different order, for instance, this way:

$$0, 1, -1, 2, -2, 3, -3, 4, \ldots.$$

More formally, if we define

$$x_i = \begin{cases} -\frac{i}{2} & \text{if } i \text{ is even} \\ \frac{i+1}{2} & \text{if } i \text{ is odd,} \end{cases}$$

then one can prove that the sequence $x_0, x_1, \ldots$ consists entirely of integers, and every integer is on it.[56]

Notice that the "work" of defining this sequence amounts to making sure that each index $i \in \mathbb{N}$ corresponds to an integer and every integer is mapped to by some such index $i \in \mathbb{N}$. The insight here is that presenting a "list" of integers is the same thing as defining a function $f \colon \mathbb{N} \to \mathbb{Z}$ that is surjective.

Exercise 156

Prove that the function $f \colon \mathbb{N} \to \mathbb{Z}$ defined by

$$f(n) = \begin{cases} -\frac{n}{2} & \text{if } n \text{ is even} \\ \frac{n+1}{2} & \text{if } n \text{ is odd,} \end{cases}$$

is surjective. Conclude that $\mathbb{Z}$ is countable.

In general, then, X is countable if and only if there is a surjection $f \colon \mathbb{N} \to X$, since in this case the sequence defined by setting $x_i = f(i)$ is a list of the elements of X in the sense defined previously.

Exercise 157

(a) Show that the union of two countable sets is countable.
(b) Show that the union of finitely many countable sets is countable.

---

[56] See Exercise 156.

(c) Show that the union of countably many countable sets is countable (i.e., if C is a countable set, each of whose elements is a countable set, then also $\bigcup_{X \in C} X$ is countable).

Exercise 158

Show that $\mathbb{Q}$ is countable.

Part of the intuition behind surjective functions $f: X \to Y$ is that there is "enough" of X to "completely cover" Y, and in this sense, X is at least a "big" as Y. From this perspective, a countable set is one that is no "bigger" than $\mathbb{N}$.[57] A similar intuition motivates the classical definition of when two infinite sets X and Y count as "the same size", namely, when there exists a *bijection* $f: X \to Y$. In this case, more officially, X and Y are called *equinumerous* and, by analogy to cardinality for finite sets, we write $|X| = |Y|$.

[57] Note well that this notion of comparative size of infinite sets is also rather counterintuitive: for example, it implies that $\mathbb{N}$ is at least as "big" as $\mathbb{Z}$, despite the fact that $\mathbb{N}$ is a strict subset of $\mathbb{Z}$!

Exercise 159

Show that equinumerosity is an equivalence relation.

We conclude with Cantor's famous "diagonalization" proof to demonstrate that some sets are *not* countable.[58]

[58] Try to figure out what the proof has to do with "diagonals". Hint: draw a picture of the list determined by f!

Theorem 86 *The powerset of* $\mathbb{N}$ *is not countable.*

Proof Suppose for contradiction that there exists a surjective function $f: \mathbb{N} \to 2^{\mathbb{N}}$. Define

$$D = \{n \in \mathbb{N} : n \notin f(n)\}.$$

Clearly $D \subseteq \mathbb{N}$, so $D \in 2^{\mathbb{N}}$. Since f is surjective, this means there is some $k \in \mathbb{N}$ such that $f(k) = D$. However, by definition of D, we know that $k \in D$ iff $k \notin f(k)$ iff $k \notin D$, a contradiction! $\quad\square$

Appendix B Selected Solutions

Exercise 1

Parse trees for (c) and (d).

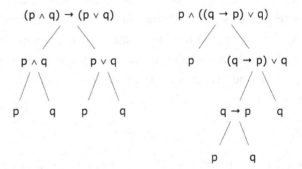

Parse trees for (e), (f), and (g).

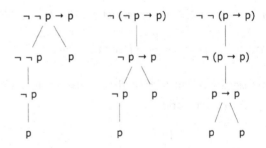

Exercise 2

For part (a), define

$$conn(p) = 0, \text{ for all } p \in \text{PROP},$$
$$conn(\neg\varphi) = conn(\varphi) + 1,$$
$$conn(\varphi * \psi) = conn(\varphi) + conn(\psi) + 1.$$

Intuitively: we set $conn(p) = 0$ because there are no connectives in the formula p; $conn(\neg\varphi) = conn(\varphi) + 1$ because the negation adds one extra connective; and $conn(\varphi * \psi) = conn(\varphi) + conn(\psi) + 1$ because the binary connective $*$, whatever it is, adds one extra connective to the total number already present in φ and ψ.

For part (b), define

$$sub(p) = \{p\}, \text{ for all } p \in \text{PROP},$$
$$sub(\neg\varphi) = sub(\varphi) \cup \{\neg\varphi\},$$
$$sub(\varphi * \psi) = sub(\varphi) \cup sub(\psi) \cup \{\varphi * \psi\}.$$

Note that here the output of the function is a *set* (of formulas) rather than a number. □

Exercise 3

Induction on φ. For the base case we have $length(p) = 1 > 0 = rank(p)$. The inductive step for $\neg$ is easy. For the binary connectives, first observe that whenever $a, b > 0$ we have $max(a, b) < a + b$.[1] Now, supposing inductively that $length(\varphi) > rank(\varphi)$ and $length(\psi) > rank(\psi)$, we have:

$$length(\varphi * \psi) = length(\varphi) + length(\psi) + 1$$
$$> rank(\varphi) + rank(\psi) + 1$$
$$> max(rank(\varphi), rank(\psi)) + 1.$$
$$= rank(\varphi * \psi).$$
□

[1] Without loss of generality assume that $a \leq b$; then $max(a, b) = b < a + b$.

Exercise 6

We prove the following by induction: for all formulas φ,

$$|sub(\varphi)| \leq 2conn(\varphi) + 1.$$

For the base case, we have $|sub(p)| = |\{p\}| = 1$ whereas $conn(p) = 0$, so $2conn(p) + 1 = 1$ as well.

Next suppose inductively the result holds for φ; then we have

$$|sub(\neg\varphi)| = |sub(\varphi) \cup \{\neg\varphi\}| = |sub(\varphi)| + 1 \leq 2conn(\varphi) + 2.$$

Since $conn(\neg\varphi) = conn(\varphi) + 1$, we also have

$$2conn(\varphi) + 2 = 2conn(\neg\varphi) < 2conn(\neg\varphi) + 1.$$

Putting these together yields $|sub(\neg\varphi)| < 2conn(\neg\varphi) + 1$, as desired.

Finally, suppose inductively the result holds for φ and ψ; then we have

$$|sub(\varphi * \psi)| = |sub(\varphi) \cup sub(\psi) \cup \{\varphi * \psi\}|$$
$$\leq |sub(\varphi)| + |sub(\psi)| + 1$$
$$\leq 2conn(\varphi) + 2conn(\psi) + 3.$$

It's also easy to see that $conn(\varphi * \psi) = conn(\varphi) + conn(\psi) + 1$, so

$$2conn(\varphi) + 2conn(\psi) + 3 = 2conn(\varphi * \psi) + 1,$$

which yields $|sub(\varphi * \psi)| \leq 2conn(\varphi * \psi) + 1$ and completes the proof.   □

Exercise 8

φ	$\leftrightarrow$	ψ	φ	$\oplus$	ψ	φ	$\mid$	ψ	φ	$\downarrow$	ψ
T	T	T	T	F	T	T	F	T	T	F	T
T	F	F	T	T	F	T	T	F	T	F	F
F	F	T	F	T	T	F	T	T	F	F	T
F	T	F	F	F	F	F	T	F	F	T	F

For binary connectives there are four combinations of input truth values, and for each of these four combinations of inputs there needs to be an output truth value of either true or false. Hence there are $2^4 = 16$ possible truth tables for binary connectives.   □

Exercise 9

Truth tables for (a) and (b).

(φ	∨	ψ)	→	φ
T	T	T	T	T
T	T	F	T	T
F	T	T	F	F
F	F	F	T	F

(φ	→	ψ)	∧	(ψ	→	φ)
T	T	T	T	T	T	T
T	F	F	F	F	T	T
F	T	T	F	T	F	F
F	T	F	T	F	T	F

Truth tables for (c) and (f).

φ	∧	φ
T	T	T
F	T	F

(φ	∧	ψ)	∨	χ
T	T	T	T	T
T	T	T	T	F
T	F	F	T	T
T	F	F	F	F
F	F	T	T	T
F	F	T	F	F
F	F	F	T	T
F	F	F	F	F

Exercise 10

Truth table for part (e).

(φ	→	ψ)	↔	(¬	(φ	∧	¬	ψ))
T	T	T	T	T	T	F	F	T
T	F	F	T	F	T	T	T	F
F	T	T	T	T	F	F	F	T
F	T	F	T	T	F	F	T	F

Truth table for part (g).

$(\varphi$	$\rightarrow$	$(\psi$	$\rightarrow$	$\chi))$	$\rightarrow$	$(\varphi$	$\rightarrow$	$\psi)$	$\rightarrow$	$(\varphi$	$\rightarrow$	$\chi))$
T	T	T	T	T	T	T	T	T	T	T	T	T
T	F	T	F	F	T	T	T	T	F	T	F	F
T	T	F	T	T	T	T	F	F	T	T	T	T
T	T	F	T	F	T	T	F	F	T	T	F	F
F	T	T	T	T	T	F	T	T	T	F	T	T
F	T	T	F	F	T	F	T	T	T	F	T	F
F	T	F	T	T	T	F	T	F	T	F	T	T
F	T	F	T	F	T	F	T	F	T	F	T	F

Exercise 12

For part (a), we proceed by induction on φ. For the base case, suppose that φ is a primitive proposition. There are two subcases to consider. First, if $\varphi = p$, then $\varphi[\psi_1/p] = \psi_1$ and $\varphi[\psi_2/p] = \psi_2$ and the result follows immediately. Otherwise, if $\varphi \neq p$, then $\varphi[\psi_1/p] = \varphi[\psi_2/p] = \varphi$, and again the result is immediate.

Now suppose inductively the result holds for φ. We want to show it also holds for $\neg\varphi$. By definition of substitution, we have $(\neg\varphi)[\psi_1/p] = \neg(\varphi[\psi_1/p])$ and $(\neg\varphi)[\psi_2/p] = \neg(\varphi[\psi_2/p])$; moreover, the inductive hypothesis tells us that $[\![\psi_1]\!]_v = [\![\psi_2]\!]_v$ implies $[\![\varphi[\psi_1/p]]\!]_v = [\![\varphi[\psi_2/p]]\!]_v$, from which of course it follows that $[\![\psi_1]\!]_v = [\![\psi_2]\!]_v$ implies $[\![\neg(\varphi[\psi_1/p])]\!]_v = [\![\neg(\varphi[\psi_2/p])]\!]_v$, which establishes the desired result.

Finally, suppose inductively the result holds for φ and φ'; we'll show it holds for $\varphi * \varphi'$ (where as usual $*$ denotes any of the binary connectives). By definition of substitution, we have $(\varphi * \varphi')[\psi_1/p] = \varphi[\psi_1/p] * \varphi'[\psi_1/p]$ and $(\varphi * \varphi')[\psi_2/p] = \varphi[\psi_2/p] * \varphi'[\psi_2/p]$. Moreover, the inductive hypothesis tells us that $[\![\psi_1]\!]_v = [\![\psi_2]\!]_v$ implies $[\![\varphi[\psi_1/p]]\!]_v = [\![\varphi[\psi_2/p]]\!]_v$ and $[\![\varphi'[\psi_1/p]]\!]_v = [\![\varphi'[\psi_2/p]]\!]_v$; it follows that $[\![\psi_1]\!]_v = [\![\psi_2]\!]_v$ implies $[\![\varphi[\psi_1/p] * \varphi'[\psi_1/p]]\!]_v = [\![\varphi[\psi_2/p] * \varphi'[\psi_2/p]]\!]_v$, which establishes the desired result.

Now for part (b), observe that if $\vDash \psi_1 \leftrightarrow \psi_2$ then, for all valuations v, we have $[\![\psi_1 \leftrightarrow \psi_2]\!]_v = true$, or equivalently, $[\![\psi_1]\!]_v = [\![\psi_2]\!]_v$. Thus, by part (a), for all valuations v we have $[\![\varphi[\psi_1/p]]\!]_v =$

$\llbracket\varphi[\psi_2/p]\rrbracket_v$, or equivalently $\llbracket\varphi[\psi_1/p]\rrbracket \leftrightarrow \varphi[\psi_2/p]\rrbracket_v = true$, which just means $\vDash \varphi[\psi_1/p] \leftrightarrow \varphi[\psi_2/p]$. □

Exercise 13

We show that the nand connective[2] is adequate. By Theorem 3, it's enough to show that every formula in $\mathcal{L}(\neg, \wedge)$ is equivalent to some formula in $\mathcal{L}(\,|\,)$. We can do this by induction on all φ in $\mathcal{L}(\neg, \wedge)$. The base case, where φ is a primitive proposition, is trivial since every primitive proposition in $\mathcal{L}(\neg, \wedge)$ is also in $\mathcal{L}(\,|\,)$.

[2] Also known as the Sheffer stroke.

Now suppose the result holds for φ. So let φ' be a formula in $\mathcal{L}(\,|\,)$ that is equivalent to φ. Then it is easy to check that $\varphi'|\varphi'$ is equivalent to $\neg\varphi$. Indeed, we have the following.

$\neg$	φ	$\leftrightarrow$	$(\varphi'$	$|$	$\varphi')$
F	T	T	T	F	T
T	F	T	F	T	F

Since $\varphi'|\varphi' \in \mathcal{L}(\,|\,)$, this completes the inductive step for $\neg$.

Finally, suppose the result holds for φ and ψ, and let φ' and ψ' be formulas in $\mathcal{L}(\,|\,)$ that are equivalent to φ and ψ, respectively. Then it is easy to check that $(\varphi'|\psi')|(\varphi'|\psi')$ is equivalent to $\varphi\wedge\psi$.[3] Indeed, we have the following.

[3] Some intuition: since "nand" is essentially the negation of "and", we should be able to recreate $\varphi \wedge \psi$ by negating $\varphi \mid \psi$. And we saw already how to simulate negation using just the Sheffer stroke!

$(\varphi$	$\wedge$	$\psi)$	$\leftrightarrow$	$((\varphi'$	$|$	$\psi')$	$|$	$(\varphi'$	$|$	$\psi'))$
T	T	T	T	T	F	T	T	T	F	T
T	F	F	T	T	T	F	F	T	T	F
F	F	T	T	F	T	T	F	F	T	T
F	F	F	T	F	T	F	F	F	T	F

Since $(\varphi'|\psi')|(\varphi'|\psi') \in \mathcal{L}(\,|\,)$, this completes the inductive step for $\wedge$. □

Exercise 14

We prove the first claim by induction on all $\varphi \in \mathcal{L}(\vee, \rightarrow)$. For the base case where φ is primitive the result is obvious, since we can simply choose v with $v(\varphi) = true$.

Now suppose inductively the result holds for φ and ψ; we'll show it holds for $\varphi \vee \psi$ and for $\varphi \to \psi$. By the inductive hypothesis, we can find a v such that $[\![\varphi]\!]_v = true$. But in this case, by definition of $\vee$, we also have $[\![\varphi \vee \psi]\!]_v = true$. Similarly, we can find a v' such that $[\![\psi]\!]_{v'} = true$, and in this case we see that $[\![\varphi \to \psi]\!]_{v'} = true$. This completes the proof of the claim.

To show that $\{\vee, \to\}$ is not adequate, consider the formula $p \wedge \neg p$. It is easy to check that for all valuations v, we have $[\![p \wedge \neg p]\!]_v = false$. But then, by the claim we just proved, we know that this formula cannot be equivalent to any formula in $\mathcal{L}(\vee, \to)$. □

Exercise 15

Consider the four combinations of truth values for any fixed pair of primitive propositions p and q, and show that each formula $\varphi \in \mathcal{L}(\neg, \leftrightarrow)$ is *true* for an even number of these combinations.

Exercise 17

Although this question is notationally heavy, the solutions are quite elementary. For example, to show that $[\![\varphi \wedge \psi]\!] = [\![\varphi]\!] \cap [\![\psi]\!]$, it suffice to observe that $v \in [\![\varphi \wedge \psi]\!]$ if and only if $v \in [\![\varphi]\!]$ and $v \in [\![\psi]\!]$. But this follows immediately from the definition of $\wedge$: $v \in [\![\varphi \wedge \psi]\!]$ iff $[\![\varphi \wedge \psi]\!]_v = true$ iff $[\![\varphi]\!]_v = true$ and $[\![\psi]\!]_v = true$ iff $v \in [\![\varphi]\!]$ and $v \in [\![\psi]\!]$. Parts (b), (c), and (d) are established similarly. For (e), the key observation is that $\vDash \varphi \to \psi$ iff every valuation that makes φ true also makes ψ true, which is exactly what is expressed by the containment $[\![\varphi]\!] \subseteq [\![\psi]\!]$.

Now, consider the map $[\![\varphi]\!] \mapsto [\varphi]$. This is well-defined, because if $[\![\varphi]\!] = [\![\varphi']\!]$, then for each valuation v, either $[\![\varphi]\!]_v = [\![\varphi']\!]_v = true$ or $[\![\varphi]\!]_v = [\![\varphi']\!]_v = false$, so φ and φ' are logically equivalent, hence $[\varphi] = [\varphi']$. In fact, it is a bijection: it is clearly surjective, and if $[\![\varphi]\!] \neq [\![\psi]\!]$, then there must be some v such that $[\![\varphi]\!]_v \neq [\![\psi]\!]_v$, so they are not logically equivalent, which implies $[\varphi] \neq [\psi]$. □

Exercise 18

For part (a), we have the following.

$$\varphi_1 = \varphi \to (\psi \to \varphi) \qquad\qquad\qquad (L1)$$
$$\varphi_2 = (\varphi \to (\psi \to \varphi)) \to ((\varphi \to \psi) \to (\varphi \to \varphi)) \quad (L2)$$
$$\varphi_3 = (\varphi \to \psi) \to (\varphi \to \varphi) \qquad\qquad (1+2 \text{ MP})$$

For part (b), we have the following.

$$\varphi_1 = \psi \to (\varphi \to \psi) \qquad\qquad\qquad (L1)$$
$$\varphi_2 = (\psi \to (\varphi \to \psi)) \to (\varphi \to (\psi \to (\varphi \to \psi))) \quad (L1)$$
$$\varphi_3 = \varphi \to (\psi \to (\varphi \to \psi)) \qquad\qquad (1+2 \text{ MP})$$

For part (d), we have the following.

$$\varphi_1 = \neg\varphi \qquad\qquad\qquad\qquad (\text{assump.})$$
$$\varphi_2 = \neg\varphi \to (\neg\psi \to \neg\varphi) \qquad (L1)$$
$$\varphi_3 = \neg\psi \to \neg\varphi \qquad\qquad (1+2 \text{ MP})$$
$$\varphi_4 = (\neg\psi \to \neg\varphi) \to (\varphi \to \psi) \quad (L3)$$
$$\varphi_5 = \varphi \to \psi \qquad\qquad\qquad (3+4 \text{ MP})$$

And for part (g), we have the following.

$$\varphi_1 = \neg(\psi \to \chi) \to \neg\varphi \qquad\qquad\qquad (\text{assump.})$$
$$\varphi_2 = (\neg(\psi \to \chi) \to \neg\varphi) \to (\varphi \to (\psi \to \chi)) \quad (L3)$$
$$\varphi_3 = \varphi \to (\psi \to \chi) \qquad\qquad\qquad (1+2 \text{ MP})$$
$$\varphi_4 = (\varphi \to (\psi \to \chi)) \to ((\varphi \to \psi) \to (\varphi \to \chi)) \quad (L2)$$
$$\varphi_5 = (\varphi \to \psi) \to (\varphi \to \chi) \qquad\qquad (3+4 \text{ MP})$$
$$\varphi_6 = \varphi \to \psi \qquad\qquad\qquad\qquad (\text{assump.})$$
$$\varphi_7 = \varphi \to \chi \qquad\qquad\qquad\qquad (5+6 \text{ MP}) \qquad \square$$

Exercise 19

Clearly every deduction is a deduction+, so one direction of the biconditional is immediate. For the other direction, we will show that for every $n \geq 1$, for every φ, if there is a deduction+ of φ of length n, then there is a deduction of φ.

We proceed by induction on n. In the base case $n = 1$, the deduction+ must be itself be a deduction, since a one-line deduction+ is either an axiom or an assumption. Now suppose inductively the result holds for each deduction+ of length less than n; we must show it holds for any deduction+ of length n. So let $\varphi_1, \ldots, \varphi_n$ be a deduction+. There are three cases to consider.

Case 1: φ_n is an axiom or an assumption. Then as before, clearly there is a deduction of φ_n, and we're done.

Case 2: φ_n follows from two previous lines φ_j and $\varphi_k = \varphi_j \to \varphi_n$ (i.e., by modus ponens). Then φ_j and φ_k each have a deduction+ that is fewer than n lines long, so by the inductive hypothesis each of them has a deduction as well. Let $\chi_1, \ldots, \chi_{n'}$ be a deduction of φ_j and $\xi_1, \ldots, \xi_{n''}$ be a deduction of φ_k; then $\chi_1, \ldots, \chi_{n'}, \xi_1, \ldots, \xi_{n''}, \varphi_n$ is a deduction of φ_n.

Case 3: $\varphi_n = \psi_1 \to \psi_3$ follows from two previous lines $\varphi_j = \psi_1 \to \psi_2$ and $\varphi_k = \psi_2 \to \psi_3$. Then as in case 2, from the induction hypothesis we obtain $\vdash \psi_1 \to \psi_2$ and $\vdash \psi_2 \to \psi_3$. From Proposition 6 and the Deduction Theorem (Theorem 8), we know that $\vdash (\psi_1 \to \psi_2) \to ((\psi_2 \to \psi_3) \to (\psi_1 \to \psi_3))$; thus, by concatenating a deduction of $\psi_1 \to \psi_2$ to a deduction of $\psi_2 \to \psi_3$ to a deduction of $(\psi_1 \to \psi_2) \to ((\psi_2 \to \psi_3) \to (\psi_1 \to \psi_3))$ and applying modus ponens twice, we can obtain a deduction of $\psi_1 \to \psi_3$, as desired. □

Exercise 20

For part (a), by the Deduction Theorem, it is enough to show that $\{\neg\psi \to \neg\varphi, \neg\psi \to \varphi\} \vdash \psi$. Observe the following.

φ_1	$=$	$\neg\psi \to \neg\varphi$	(assump.)
φ_2	$=$	$(\neg\psi \to \neg\varphi) \to (\varphi \to \psi)$	(L3)
φ_3	$=$	$\varphi \to \psi$	(1+2 MP)
φ_4	$=$	$\neg\psi \to \varphi$	(assump.)
φ_5	$=$	$\neg\psi \to \psi$	(3+4 HS)

This shows that $\{\neg\psi \to \neg\varphi, \neg\psi \to \varphi\} \vdash \neg\psi \to \psi$. By Proposition 9, we also have $\vdash (\neg\psi \to \psi) \to \psi$, from which the result easily follows.

For (b), first note that the proof of the Deduction Theorem does not use (L3); it follows that the Deduction Theorem also holds for ⊩. Thus, it suffices to prove that $\{\neg\psi \rightarrow \neg\varphi, \varphi\} \Vdash \psi$. We have the following.

$$
\begin{array}{lll}
\varphi_1 & = & \neg\psi \rightarrow \neg\varphi & \text{(assump.)} \\
\varphi_2 & = & (\neg\psi \rightarrow \neg\varphi) \rightarrow ((\neg\psi \rightarrow \varphi) \rightarrow \psi) & \text{(L3')} \\
\varphi_3 & = & (\neg\psi \rightarrow \varphi) \rightarrow \psi & \text{(1+2 MP)} \\
\varphi_4 & = & \varphi \rightarrow (\neg\psi \rightarrow \varphi) & \text{(L1)} \\
\varphi_5 & = & \varphi & \text{(assump.)} \\
\varphi_6 & = & \neg\psi \rightarrow \varphi & \text{(4+5 MP)} \\
\varphi_7 & = & \psi & \text{(3+6 MP)}
\end{array}
$$

Now for (c), let's prove that $\vdash \varphi$ implies $\Vdash \varphi$ (the proof of the reverse implication is completely analogous). We'll proceed by induction on the length of the deduction of φ from (L1), (L2), and (L3); that is, we'll prove the following: for every n, for every φ, if there is a deduction of φ from (L1), (L2), and (L3) of length n, then there is a deduction of φ from (L1), (L2), and (L3').

For the base case $n = 1$ we know that φ is an axiom. If φ is an instance of (L1) or (L2), then clearly $\Vdash \varphi$. If φ is an instance of (L3), then we know that $\Vdash \varphi$ from part (b).

For the inductive step, assume that every formula with a deduction from (L1), (L2), and (L3) fewer than n lines long is deducible from (L1), (L2), and (L3'); we will show that if φ has a deduction from (L1), (L2), and (L3) that is n lines long, then $\Vdash \varphi$. If φ is an axiom, then we can reason as in the base case. So assume that φ follows by modus ponens from two previous lines φ_j and φ_k. Then by the induction hypothesis, since φ_j and φ_k have deductions from (L1), (L2), and (L3) fewer than n lines long, we know that $\Vdash \varphi_j$ and $\Vdash \varphi_k$. Thus, putting these deductions together and applying (MP), we obtain $\Vdash \varphi$, as desired. □

Exercise 21

This is a simple proof by contradiction. If $p \rightarrow (p \rightarrow q)$ were deducible, then by soundness we could conclude it is a tautology. However, it is easy to see that this formula is not a tautology: consider a valuation v such that $v(p) = true$ and $v(q) = false$. □

Exercise 22

As above, use soundness to derive a contradiction!

Exercise 24

Suppose that Γ is inconsistent; we want to show that for any φ, $\Gamma \vdash \varphi$. We know there is some ψ such that $\Gamma \vdash \psi$ and $\Gamma \vdash \neg\psi$. Let φ be any formula. From Exercise 18(e), we know that $\vdash \neg\psi \rightarrow (\psi \rightarrow \varphi)$, hence also $\Gamma \vdash \neg\psi \rightarrow (\psi \rightarrow \varphi)$. Putting these three deductions from Γ together (and applying modus ponens twice) shows that $\Gamma \vdash \varphi$, as desired.

For an alternative proof, consider the contrapositive of Proposition 11, which says that if $\Gamma \cup \{\neg\varphi\}$ is inconsistent, then $\Gamma \vdash \varphi$. Of course, if Γ itself is inconsistent, then it is easy to see that $\Gamma \cup \{\neg\varphi\}$ is also inconsistent, so again we obtain the desired result.   □

Exercise 25

Suppose for contradiction that $\varphi \notin \Gamma$. Then, by maximality, we know we must have $\neg\varphi \in \Gamma$. But then we clearly have $\Gamma \vdash \neg\varphi$, which implies (since $\Gamma \vdash \varphi$) that Γ is inconsistent, a contradiction.   □

Exercise 27

For strong soundness, we proceed by induction on the length of the deduction, as in the proof of Theorem 10. Specifically, we prove the following: for all $n \geq 1$, if $\varphi_1, \ldots, \varphi_n$ is a deduction from Γ, then $\Gamma \vDash \varphi_n$.

For the base case when $n = 1$, we have a deduction φ_1 from Γ that's only one line long. Thus either φ_1 is an instance of an axiom or else $\varphi_1 \in \Gamma$. In either case, it is easy to see that $\Gamma \vDash \varphi_1$.

Now suppose inductively that the result holds for all $k \leq n$, and let $\varphi_1, \ldots, \varphi_{n+1}$ be a deduction from Γ. If φ_{n+1} is an axiom or in Γ we can reason as before to conclude that $\Gamma \vDash \varphi_{n+1}$, so suppose instead that φ_{n+1} follows by modus ponens from two previous lines φ_i and φ_j, where $\varphi_i = \varphi_j \rightarrow \varphi_{n+1}$. Then by the inductive hypothesis we have $\Gamma \vDash \varphi_j$ and $\Gamma \vDash \varphi_j \rightarrow \varphi_{n+1}$. But from this it follows that $\Gamma \vDash \varphi_{n+1}$: for if not, then there is some v that satisfies Γ but with $[\![\varphi_{n+1}]\!]_v = \text{false}$, so since v satisfies Γ and $\Gamma \vDash \varphi_j$, we know $[\![\varphi_j]\!]_v = \text{true}$, which implies that $[\![\varphi_j \rightarrow \varphi_{n+1}]\!]_v = \text{false}$, contradicting $\Gamma \vDash \varphi_j \rightarrow \varphi_{n+1}$.[4]

For strong completeness, we prove the contrapositive. Assume that $\Gamma \nvdash \varphi$. Then $\Gamma \cup \{\neg\varphi\}$ is consistent by Proposition 11. By Lindenbaum's Lemma (Lemma 13), there is a maximally consistent $\Gamma' \supseteq \Gamma \cup \{\neg\varphi\}$. Therefore, by the Truth Lemma (Lemma 14), we know that $v_{\Gamma'}$ satisfies Γ and $[\![\neg\varphi]\!]_{v_{\Gamma'}} = \text{true}$, so $v_{\Gamma'}$ witnesses that $\Gamma \nvDash \varphi$.[5]   □

[4] Notice that this argument is almost identical to that given in the proof of Theorem 10.

[5] Again, notice how similar this argument is to the proof of regular completeness—the tools and lemmas we developed turn out to be powerful enough to reach this stronger result with almost no extra work.

Exercise 28

First suppose that Γ is consistent. Then by Lindenbaum's Lemma (Lemma 13) there is a maximally consistent $\Gamma' \supseteq \Gamma$, and by the Truth Lemma (Lemma 14) we know that $v_{\Gamma'}$ satisfies Γ.

For the converse, suppose that Γ is not consistent. Then by definition there is a formula ψ such that $\Gamma \vdash \psi$ and $\Gamma \vdash \neg\psi$. By strong soundness (Exercise 27), we therefore have $\Gamma \vDash \psi$ and $\Gamma \vDash \neg\psi$. Since no valuation makes both ψ and $\neg\psi$ true, it follows that Γ must not be satisfiable.   □

Exercise 29

First the easy direction: clearly if Γ is satisfiable then every finite subset is also satisfiable. For the converse, first observe that by Exercise 28, it is equivalent to prove that if every finite subset of Γ is *consistent* then Γ must also be consistent. We will prove the contrapositive.

Suppose that Γ is not consistent. Let ψ be such that $\Gamma \vdash \psi$ and $\Gamma \vdash \neg\psi$, and let $\psi_1, \ldots, \psi_n$ and $\psi'_1, \ldots, \psi'_m$ be deductions of ψ and $\neg\psi$ from Γ, respectively. Let

$$F = \Gamma \cap \{\psi_1, \ldots, \psi_n, \psi'_1, \ldots, \psi'_m\}.$$

Clearly F is a finite subset of Γ, and also $F \vdash \psi$ and $F \vdash \neg\psi$ (using the very same deductions). Thus, not every finite subset of Γ is consistent.   □

Exercise 31

(a) By Exercise 28, it is enough to show that P_v is satisfiable. But of course it is: v satisfies P_v.

(b) By part (a), Lindenbaum's Lemma (Lemma 13) guarantees *at least one* maximally consistent extension of P_v, so we need to show there cannot be more than one. For this a small lemma is useful: *If Γ and Γ' are maximally consistent sets and $\Gamma \cap \text{PROP} = \Gamma' \cap \text{PROP}$, then $\Gamma = \Gamma'$.* To see this, first observe that since Γ and Γ' contain exactly the same primitive propositions, we must have $v_\Gamma = v_{\Gamma'}$. Now consider any formula φ, and notice that by two applications of the Truth Lemma (Lemma 14) we have

$$\varphi \in \Gamma \text{ iff } [\![\varphi]\!]_{v_\Gamma} = true$$
$$\text{iff } [\![\varphi]\!]_{v_{\Gamma'}} = true$$
$$\text{iff } \varphi \in \Gamma'.$$

This proves our lemma. Now we can apply it: suppose that Γ and Γ' are both extensions of P_v. Then clearly $\Gamma \cap \text{PROP} = v^{-1}(true) = \Gamma' \cap \text{PROP}$, so our lemma guarantees that $\Gamma = \Gamma'$.

(c) First note that by definition Γ_{v_Γ} is an extension of P_{v_Γ}, so

$$\Gamma_{v_\Gamma} \cap \text{PROP} = v_\Gamma^{-1}(true) = \Gamma \cap \text{PROP}.$$

Thus, by our lemma from part (b), we have $\Gamma_{v_\Gamma} = \Gamma$. For the other equality, observe that for each primitive proposition p we have

$$v_{\Gamma_v}(p) = true \text{ iff } p \in \Gamma_v$$
$$\text{iff } p \in P_v$$
$$\text{iff } v(p) = true,$$

which shows that $v_{\Gamma_v} = v$.

(d) What we have shown in (c) is that the two functions mapping valuations to maximally consistent sets ($v \mapsto \Gamma_v$) and mapping maximal consistent sets to valuations ($\Gamma \mapsto v_\Gamma$)

are in fact inverses of each other. This means that there is a bijective correspondence between valuations and maximally consistent sets.

(e) By part (d), we know that there are as many maximally consistent sets as there are valuation functions. Since each valuation function is a map from a countable set (namely, PROP) to a set of size 2 (namely, {*true*, *false*}), it follows that the number of valuation functions is the same as the number of functions from $\mathbb{N}$ to $\{0, 1\}$. The set of all such functions is sometimes denoted $2^{\mathbb{N}}$; it is well known to have size continuum—it encodes, for example, all binary expansions of real numbers between 0 and 1. □

Exercise 32

(a) $\forall x(x = \bar{0} \lor \ell(\bar{0}, x))$.

(b) $\exists x \forall y(y = x \lor \ell(x, y))$.

(c) $\neg \exists x \forall y(y = x \lor \ell(y, x))$.

(d) $\forall x \exists y(\ell(x, y) \land \forall z \neg(\ell(x, z) \land \ell(z, y)))$.

(e) $\forall x \exists y(\neg(y = x) \land \ell(y, x))$.

(f) $\forall x(\exists y(x = m(y, y)) \to \ell(\bar{0}, x))$.

(g) Here, it's helpful to define some abbreviations: let $e(x)$ be an abbreviation for $\exists y(x = a(y, y))$, and let $p(x)$ abbreviate

$$\ell(\bar{1}, x) \land \forall z \forall z'(x = m(z, z') \to (z = \bar{1} \lor z' = \bar{1})).$$

Then the solution is:

$$\exists x(e(x) \land p(x) \land \forall w((e(w) \land p(w)) \to w = x)).$$

(h) $\forall y \forall y' \exists x(p(x) \land ((\ell(m(y, y), x) \land \ell(x, m(y', y'))) \lor (\ell(m(y', y'), x) \land \ell(x, m(y, y)))))$. □

Exercise 33

For each variable x_i, recursively define $f^0(x_i) = x_i$, and for $k \in \mathbb{N}$, $f^{k+1}(x_i) = f(f^k(x_i))$. Then the set of terms is simply

$$\{f^k(x_i) : k \in \mathbb{N} \text{ and } x_i \text{ is a variable}\}.$$

Exercise 34

One might define $BV(\varphi)$ recursively as follows:

- $BV(\bar{c}) = \varnothing$
- $BV(x) = \varnothing$
- $BV(f(t_1, \ldots, t_n)) = \varnothing$
- $BV(r(t_1, \ldots, t_n)) = \varnothing$
- $BV(t_1 = t_2) = \varnothing$
- $BV(\neg\varphi) = BV(\varphi)$
- $BV(\varphi \to \psi) = BV(\varphi) \cup BV(\psi)$
- $BV(\forall x\varphi) = BV(\exists x\varphi) = BV(\varphi) \cup \{x\}$.

Notice, however, that according to this definition, $BV(\varphi)$ includes even those variables that *only* occur attached to a quantifier, for example, $BV(\forall x(\bar{c})) = \{x\}$. One might reasonably wish to instead count only those variables that actually occur *as terms* in the formula. Of course, we can also recursively define this notion of a variable "occurring as a term":

- $TV(\bar{c}) = \varnothing$
- $TV(x) = \{x\}$
- $TV(f(t_1, \ldots, t_n)) = TV(t_1) \cup \cdots \cup TV(t_n)$
- $TV(r(t_1, \ldots, t_n)) = TV(t_1) \cup \cdots \cup TV(t_n)$
- $TV(t_1 = t_2) = TV(t_1) \cup TV(t_2)$
- $TV(\neg\varphi) = TV(\varphi)$
- $TV(\varphi \to \psi) = TV(\varphi) \cup TV(\psi)$
- $TV(\forall x\varphi) = TV(\exists x\varphi) = TV(\varphi)$.

The alternative notion of "bound variables" suggested above would thus be given by $BV(\varphi) \cap TV(\varphi)$.

It's easy to see that $FV(\varphi) \cap BV(\varphi)$ need not be empty. For instance, the formula $\forall x(x = x) \to x = x$ includes both free and bound instances of the variable x.   □

Exercise 36

Consider the structure $\mathcal{B}$ which is just like $\mathcal{A}$ except $[\![f]\!]_\mathcal{B} = \cdot$ (i.e., multiplication). Then, unravelling the recursive definition of truth as in Proposition 17, we have $[\![\varphi]\!]_{\mathcal{B}}$ = *true* iff for all $n, m \in \mathbb{Z}$, if $n \cdot m < 0$ then $n < m$. This is clearly not the case (e.g., take $n = 1$ and $m = -1$), hence $[\![\varphi]\!]_{\mathcal{B}} = \textit{false}$, as desired. $\qquad\square$

Exercise 37

(a) $[\![\forall x(r(f(\bar{c}, x), \bar{c}))]\!]_{\mathcal{A}}$ = *true* iff for all $n \in \mathbb{Z}$, $0 - n < 0$, which is not the case (e.g., take $n = -1$). Thus $[\![\forall x(r(f(\bar{c}, x), \bar{c}))]\!]_{\mathcal{A}}$ = *false*.

(b) $[\![\forall x \forall y \forall z(r(x, y) \to r(f(x, z), f(y, z)))]\!]_{\mathcal{A}}$ = *true* iff for all $n, m, k \in \mathbb{Z}$, if $n < m$ then $n - k < m - k$, which is true.

(c) $[\![\forall x \exists y(f(x, y) = \bar{c})]\!]_{\mathcal{A}}$ = *true* iff for all $n \in \mathbb{Z}$, there is an $m \in \mathbb{Z}$ such that $n - m = 0$, which is true (take $m = n$).

(d) $[\![\exists y \forall x(f(x, y) = \bar{c})]\!]_{\mathcal{A}}$ = *true* iff there is an $n \in \mathbb{Z}$ such that, for all $m \in \mathbb{Z}$, $m - n = 0$. This is false since no matter what n is, we can find an m such that $m - n \neq 0$ (e.g., take $m = n + 1$). $\qquad\square$

Exercise 38

(a) It is easy to see that $[\![s(s(s(s(s(\bar{0})))))]\!]_{\mathcal{N}}$ = 5, and, for instance, also $[\![a(s(s(s(s(s(\bar{0}))))), \bar{0})]\!]_{\mathcal{N}}$ = 5.

(b) Recursively define $s^0(\bar{0}) = \bar{0}$, and for $k \in \mathbb{N}$, $s^{k+1}(\bar{0}) = s(s^k(\bar{0}))$. Then it is easy to see that $[\![s^n(\bar{0})]\!]_{\mathcal{N}} = n$.

(c) Given any term t, recursively define $g^0(t) = t$, and for $k \in \mathbb{N}$, $g^{k+1}(t) = a(g^k(t), \bar{0})$. Then it is easy to see that for all k, $[\![g^k(s^n(x))]\!]_{\mathcal{N}} = n$, and moreover that all of the terms in $\{g^k(s^n(x)) : k, n \in \mathbb{N}\}$ are distinct. $\qquad\square$

Exercise 39

(a) $[\![\forall x(m(x, s(\bar{0})) = x)]\!]_{\mathcal{N}}$ = *true* iff for all $n \in \mathbb{N}$, $n \cdot 1 = n$, which is true.

(b) $[\![\forall x \forall y(a(x, \bar{0}) = y \to a(y, \bar{0}) = x)]\!]_{\mathcal{N}}$ = *true* iff for all $n, m \in \mathbb{N}$, $n + 0 = m$ implies $m + 0 = n$, which is true.

(c) $[\![\forall x \forall y \exists z(a(x, z) = y)]\!]_{\mathcal{N}}$ = *true* iff for all $n, m \in \mathbb{N}$, there is a $k \in \mathbb{N}$ such that $n + k = m$, which is false (it fails whenever $m < n$).

[6] Notice that the symbol "m" is overloaded here—it stands for the function symbol that is interpreted as multiplication in the model $\mathcal{N}$, but we have also subsequently introduced it to stand for some arbitrary natural number. This is bad practice in general as it can create ambiguity or confusion.

(e) $[\![\forall x \forall y (m(x, s(y)) = a(m(x,y), x))]\!]_{\mathcal{N}} = true$ iff for all $n, m \in \mathbb{N}$,[6] $n \cdot (m + 1) = n \cdot m + n$, which is true.

(f) $[\![\exists x (a(s(x), s(x)) = m(s(x), s(x)))]\!]_{\mathcal{N}} = true$ iff for some $n \in \mathbb{N}$, $(n + 1) + (n + 1) = (n + 1) \cdot (n + 1)$, which is true (take $n = 1$). □

Exercise 40

The solution here is actually almost trivial—the hard part is parsing the statement! Admittedly, it looks pretty strange. First note that since t is a closed term, we know that $[\![t]\!]_A$ must be an element in the domain of discourse, A. That's why we can put a tilde over it.

Now for all $a \in A$, we know by definition that $[\![\tilde{a}]\!]_A = a$; in particular, this means that $[\![\widetilde{[\![t]\!]_A}]\!]_A = [\![t]\!]_A$. Thus, applying the recursive definition of truth, we obtain:

$$[\![t = \widetilde{[\![t]\!]_A}]\!]_A = true \quad \text{iff} \quad [\![t]\!]_A = [\![\widetilde{[\![t]\!]_A}]\!]_A$$
$$\text{iff} \quad [\![t]\!]_A = [\![t]\!]_A,$$

which of course is true. □

Exercise 41

Let $\mathcal{A} = (A, P, Q, R)$ be an arbitrary S-structure.

(a) We have $[\![\forall x r(x, x) \rightarrow \exists y \forall x r(x, y)]\!]_A = true$ iff the following statement holds:

if for all $a \in A$, $(a, a) \in R$,

then there is a $b \in A$ such that for all $c \in A$, $(c, b) \in R$.

This is not true in general; for example, it fails when A is the set of all people and $R(x, y)$ means "x and y have the same birthday"—it is of course true that all people have the same birthday as themselves, but it does not follow that there is someone (b) that has the same birthday as everyone else. Therefore the given formula is not valid.

(b) We have $[\![\forall x \exists y r(x, y) \rightarrow \exists y \forall x r(x, y)]\!]_A = true$ iff the following statement holds:

if for all $a \in A$ there is a $b \in A$ such that $(a, b) \in R$, then there is a $c \in A$ such that, for all $d \in A$, $(d, c) \in R$.

This is not true in general; for example, it fails when $A = \mathbb{N}$ and $R = \{(n, m) : n < m\}$. In this case, the antecedent holds because every number is smaller than some other number, but the consequent fails because there is not a number that is bigger than all the numbers. Thus the given formula is not valid.

(c) We have $[\![\exists y \forall x r(x, y) \to \forall x \exists y r(x, y)]\!]_A = true$ iff the following statement holds:

> if there is an $a \in A$ such that for all $b \in A$, $(b, a) \in R$,
>
> then for all $c \in A$, there exists a $d \in A$ such that $(c, d) \in R$.

This is true. If a has the special property that all elements are related by R to it, it follows that we can take $d = a$ to satisfy the above consequent. Since A was arbitrary, it follows that the given formula is valid.

(g) We have $[\![\forall x p(x) \to \forall y p(y)]\!]_A = true$ iff the following statement holds:

> if for all $a \in A$ we have $a \in P$,
>
> then for all $b \in A$ we have $b \in P$.

This is clearly true: the fact that every $a \in A$ is in P immediately tells us that every $b \in A$ is in P (the letter we use is irrelevant).

(k) We have $[\![\forall y (\forall x p(x) \to p(y))]\!]_A = true$ iff the following statement holds:

> for all $a \in A$ we have: if for all $b \in A$, $b \in P$, then $a \in P$.

This is clearly true, since the antecedent of the conditional is just the statement that $A \subseteq P$, in which case it certainly follows that $a \in P$.

(l) We have $[\![\forall y (p(y) \to \forall x p(x))]\!]_A = true$ iff the following statement holds:

> for all $a \in A$ we have: if $a \in P$, then for all $b \in A$, $b \in P$.

This does not hold in general; for example, if $A = \mathbb{N}$ and P is the set of prime numbers, then it fails for $a = 3$ (since a is prime, but not all numbers are prime).   □

Exercise 45

For part (a), we have the following.

φ_1: $x = x$ (E1)

φ_2: $\forall x(x = x)$ (1 Gen)

For part (b), we have the following.

φ_1: $x = y \to (x = x \to y = x)$ (E3)

φ_2: $(x = y \to (x = x \to y = x)) \to ((x = y \to x = x)$

$\qquad \to (x = y \to y = x))$ (L2)

φ_3: $(x = y \to x = x) \to (x = y \to y = x)$ (1,2 MP)

φ_4: $x = x \to (x = y \to x = x)$ (L1)

φ_5: $x = x$ (E1)

φ_6: $x = y \to x = x$ (4,5 MP)

φ_7: $x = y \to y = x$ (3,6 MP)

φ_8: $\forall y(x = y \to y = x)$ (7 Gen)

φ_9: $\forall x \forall y(x = y \to y = x)$ (8 Gen)

And for part (c), we have the following.

φ_1: $x = y \to (x = x \to y = x)$ (E3)

φ_2: $(x = y \to (x = x \to y = x)) \to ((x = y$

$\qquad \to x = x) \to (x = y \to y = x))$ (L2)

φ_3: $(x = y \to x = x) \to (x = y \to y = x)$ (1,2 MP)

φ_4: $x = x \to (x = y \to x = x)$ (L1)

φ_5: $x = x$ (E1)

φ_6: $x = y \to x = x$ (4,5 MP)

φ_7: $x = y \to y = x$ (3,6 MP)

φ_8: $y = x \to (y = z \to x = z)$ (E3)

φ_9: $x = y \to (y = z \to x = z)$ (7,8 HS)

φ_{10}: $\forall z(x = y \to (y = z \to x = z))$ (9 Gen)

φ_{11}: $\forall y \forall z(x = y \rightarrow (y = z \rightarrow x = z))$ (10 Gen)

φ_{12}: $\forall x \forall y \forall z(x = y \rightarrow (y = z \rightarrow x = z))$ (11 Gen) $\square$

Exercise 46

To say that (E1) is true in $\mathcal{A}$ is to say that $\mathcal{A} \vDash x = x$, which in turn means that $\mathcal{A} \vDash \forall x(x = x)$. This holds if and only if, for all $n \in \mathbb{Z}$, we have $\mathcal{A} \vDash \tilde{n} = \tilde{n}$, which by our new definition holds iff for all $n \in \mathbb{Z}$, $n - n$ is even, which is clearly true since $n - n = 0$ and 0 is even.

For (E2), since there is only one function symbol and it has arity 2, we have two relevant instances:[7]

$$x = y \rightarrow x + z = y + z,$$

and

$$x = y \rightarrow z + x = z + y.$$

We show the first is true in $\mathcal{A}$; the second is analogous. We have:

$$\mathcal{A} \vDash x = y \rightarrow x + z = y + z \text{ iff } \mathcal{A} \vDash \forall x \forall y \forall z(x = y \rightarrow x + z = y + z)$$

$$\text{iff} \quad \text{for all } n, m, k \in \mathbb{Z}, \mathcal{A} \vDash \tilde{n} = \tilde{m}$$

$$\rightarrow \tilde{n} + \tilde{k} = \tilde{m} + \tilde{k}$$

$$\text{iff} \quad \text{for all } n, m, k \in \mathbb{Z}, \text{ if } n - m \text{ is even,}$$

$$\text{then } n + k - (m + k) \text{ is even.}$$

And this is obviously true since $n + k - (m + k) = n - m$.

Finally, for (E3), since there are no relation symbols in the language, the only instances involve the equality symbol itself:

$$x = y \rightarrow (x = z \rightarrow y = z),$$

and

$$x = y \rightarrow (z = x \rightarrow z = y).$$

We have:

$$\mathcal{A} \vDash x = y \rightarrow (x = z \rightarrow y = z) \text{ iff } \mathcal{A} \vDash \forall x \forall y \forall z(x = y \rightarrow (x = z \rightarrow y = z))$$

$$\text{iff} \quad \text{for all } n, m, k \in \mathbb{Z}, \mathcal{A} \vDash \tilde{n} = \tilde{m}$$

$$\rightarrow (\tilde{n} = \tilde{k} \rightarrow \tilde{m} = \tilde{k})$$

[7] Technically, there are infinitely many instances of (E2) corresponding to the infinitely many different variables we could use. But it's easy to see that they are all handled the same way as these.

$$\text{iff} \quad \text{for all } n, m, k \in \mathbb{Z}, \text{ if } n - m \text{ is even,}$$
$$\text{then if } n - k \text{ is even, } m - k \text{ is even.}$$

Since the difference of any two even numbers is even, and $n - k - (n-m) = m-k$, this is true. The other instance of (E3) is shown similarly.

For part (b), define

$$F(n) = \begin{cases} n/2 & \text{if } n \text{ is even} \\ n & \text{otherwise.} \end{cases}$$

(Note that this is well defined as a function from $\mathbb{Z}$ to itself, since any even number when divided by 2 produces an integer.) In this case, an instance of (E2) is

$$x = y \rightarrow f(x) = f(y),$$

but this is not true in $\mathcal{A}'$, since for example we have $\mathcal{A}' \vDash \tilde{4} = \tilde{2}$ (since $4-2$ is even), but $\mathcal{A}' \nvDash f(\tilde{4}) = f(\tilde{2})$, since $F(4)-F(2) = 2-1 = 1$ is not even. $\qquad\qquad\square$

Exercise 48

First, observe that the sequence

$$\begin{array}{lll} \varphi_1: & \forall x \varphi(x) & \text{(assump)} \\ \varphi_2: & \varphi(y) & \text{(L4)} \\ \varphi_3: & \forall y \varphi(y) & \text{(2 Gen)} \end{array}$$

is a deduction—the second line is a legitimate instance of (L4) because, by assumption, y does not occur in $\varphi(x)$, and thus y must be free for x in φ. This shows that $\{\forall x \varphi(x)\} \vdash \forall y \varphi(y)$. We can then apply the Deduction Theorem (Theorem 20)[8] to conclude that $\vdash \forall x \varphi(x) \rightarrow \forall y \varphi(y)$.

For the converse, observe that the sequence

$$\begin{array}{lll} \psi_1: & \forall y \varphi(y) & \text{(assump)} \\ \psi_2: & \varphi(x) & \text{(L4)} \\ \psi_3: & \forall x \varphi(x) & \text{(2 Gen)} \end{array}$$

[8] Since again, by assumption, y does not occur free in $\forall x \varphi(x)$.

is a deduction—here, the second line is legitimate because we know that no occurrence of y in $\varphi(y)$ can occur within the scope of a $\forall x$ quantifier, since if it did we never would have substituted that occurrence of y for x in $\varphi(x)$. This shows that $\{\forall y \varphi(y)\} \vdash \forall x \varphi(x)$. We can then apply the Deduction Theorem once again[9] to conclude that $\vdash \forall x \varphi(x) \to \forall y \varphi(y)$. □

[9] Since x cannot occur free in $\forall y \varphi(y)$, otherwise it would also be free in $\varphi(y)$, contradicting the fact that $\varphi(y) = \varphi[y/x]$.

Exercise 52

By soundness (Theorem 26), proving that $\nvDash \forall x \exists y r(x, y) \to \exists y r(y, y)$ is enough to establish (a). And this simply requires a countermodel: for instance, consider the model $\mathcal{N}$ where the domain of discourse is $\mathbb{N}$ and $[\![r]\!]_{\mathcal{N}} = {<}$. In this case, we have $\mathcal{N} \vDash \forall x \exists y r(x, y)$ since this evaluates to the claim that for all $n \in \mathbb{N}$ there exists an $m \in \mathbb{N}$ such that $n < m$ (which is obviously true), whereas $\mathcal{N} \nvDash \exists y r(y, y)$, since this evaluates to the claim that there exists an $n \in \mathbb{N}$ such that $n < n$ (which is obviously false).

The argument presented in the exercise is flawed because the second line of the purported deduction is not actually an instance of (L4), since y is not free for x in $\exists y r(x, y)$. □

Exercise 53

The answer is no. Since $\forall x \varphi(x) \to \psi$ is closed, by the Deduction Theorem (Theorem 20) we know that

$$\{\forall x \varphi(x) \to \psi\} \vdash \forall x (\varphi(x) \to \psi)$$

is equivalent to

$$\vdash (\forall x \varphi(x) \to \psi) \to \forall x (\varphi(x) \to \psi).$$

So by soundness (Theorem 26), it is enough to show that

$$\nvDash (\forall x \varphi(x) \to \psi) \to \forall x (\varphi(x) \to \psi).$$

Suppose that $\varphi(x)$ is $p(x)$ (where p is a unary predicate symbol) and ψ is $\neg(\bar{c} = \bar{c})$, and consider the interpretation of the above formula in a structure $\mathcal{N}$ whose domain of discourse is $\mathbb{N}$ and where $p(x)$ is interpreted as "x is prime", and $\bar{c}$ is interpreted as 0. Then we have $\mathcal{N} \vDash \forall x \varphi(x) \rightarrow \psi$ because the antecedent of this conditional comes out as false (not all numbers are prime), whereas $\mathcal{N} \nvDash \forall x(\varphi(x) \rightarrow \psi)$ because the implication does not hold for all natural numbers; for example, 3 is prime but $0 = 0$. □

Exercise 55

Suppose for contradiction that Γ is not consistent over $\mathcal{S}'$, and consider deductions from Γ of ψ and $\neg\psi$. Try replacing whatever new constants appear in these deductions over $\mathcal{S}'$ with "fresh" variables (i.e., variables that appear nowhere else in the deductions). Are the results still deductions? □

Exercise 57

What does it mean for a function to be well-defined on a quotient? It means that whenever two inputs which are equal (in the quotient) are given to the function, the outputs are equal (in the quotient). In this case, then, we must show that whenever $([t_1], \ldots, [t_n]) = ([t'_1], \ldots, [t'_n])$, it follows that $[\![f]\!]_A([t_1], \ldots, [t_n]) = [\![f]\!]_A([t'_1], \ldots, [t'_n])$.

We have that $([t_1], \ldots, [t_n]) = ([t'_1], \ldots, [t'_n])$ if and only if, for each i, $[t_i] = [t'_i]$; of course $[t_i] = [t'_i]$ iff $t_i \sim t'_i$, which is equivalent to $t_i = t'_i \in \Gamma'$. Similarly, since by definition we have $[\![f]\!]_A([t_1], \ldots, [t_n]) = [f(t_1, \ldots, t_n)]$, we can see that $[\![f]\!]_A([t_1], \ldots, [t_n]) = [\![f]\!]_A([t'_1], \ldots, [t'_n])$ if and only if $[f(t_1, \ldots, t_n)] = [f(t'_1, \ldots, t'_n)]$.

Our task is thus to show that if, for each i, $t_i = t'_i$ is in Γ', it follows that $[f(t_1, \ldots, t_n)] = [f(t'_1, \ldots, t'_n)]$. To do this we make use of axiom (E2). One instance of (E2) is the following:

$$x_i = y \rightarrow f(x_1, \ldots, x_i, \ldots, x_n) = f(x_1, \ldots, y, \ldots, x_n).$$

It follows (by repeatedly applying the generalization rule of inference) that

$$\vdash \forall y \forall x_1 \cdots \forall x_n (x_i = y \rightarrow f(x_1, \ldots, x_i, \ldots, x_n) = f(x_1, \ldots, y, \ldots, x_n)).$$

Thus, through repeated uses of (L4) and modus ponens, we can obtain

$$\vdash t_i = t'_i \rightarrow f(t'_1,\ldots,t'_{i-1},t_i,t_{i+1},\ldots,t_n) = f(t'_1,\ldots,t'_{i-1},t'_i,t_{i+1},\ldots,t_n).$$

Therefore, since we know that $t_i = t'_i \in \Gamma'$ for each i, and that Γ' is maximally consistent, it must be that $f(t'_1,\ldots,t'_{i-1},t_i,t_{i+1},\ldots,t_n) = f(t'_1,\ldots,t'_{i-1},t'_i,t_{i+1},\ldots,t_n) \in \Gamma'$ as well, for each i. Hence for each i we know that

$$[f(t'_1,\ldots,t'_{i-1},t_i,t_{i+1},\ldots,t_n)] = [f(t'_1,\ldots,t'_{i-1},t'_i,t_{i+1},\ldots,t_n)],$$

and so we have

$$[f(t_1,\ldots,t_n)] = [f(t'_1,t_2,\ldots,t_n)] = [f(t'_1,t'_2,t_3,\ldots,t_n)]$$
$$= \cdots = [f(t'_1,\ldots,t'_n)],$$

as desired. □

Exercise 60

All of these are analogous to the corresponding arguments we saw for propositional logic.

For (a), strong soundness is simply proved by induction on the length of the deduction (as the proof went for soundness), taking into account the additional base case where a one-line deduction may simply consist in a formula from Γ. Strong completeness proceeds as follows: suppose that $\Gamma \nvdash \varphi$; then $\Gamma \cup \{\neg\varphi\}$ is consistent, so by the Model Existence Lemma (Lemma 27) it has a model; clearly this model does not satisfy φ, so $\nvDash \varphi$.

For (b), consistency implies satisfiability because that is exactly what the Model Existence Lemma tells us. On the other hand, if Γ is not consistent, then we have $\Gamma \vdash \varphi$ and $\Gamma \vdash \neg\varphi$ for some (in fact all) φ; then by strong soundness we have $\Gamma \vDash \varphi$ and $\Gamma \vDash \neg\varphi$, but it is impossible for any model to satisfy both φ and $\neg\varphi$, hence there must be no models that satisfy Γ.

Finally, for (c), the left-to-right direction is trivial, whereas by (b), proving the right-to-left direction is equivalent to proving that Γ is consistent whenever each of its finite subsets are. But this follows from the finitistic nature of deductions: if Γ is not consistent, then there is a deduction of a contradiction from some finite subset of Γ, which of course means that finite subset is not consistent.   □

Exercise 63

For (b), ask yourself this: if everyone already knows something that's announced, does that necessarily mean that no one learns anything from the announcement? Then read Chapter 4!

Exercise 72

(a) If $x \vDash \Box\varphi$ then for all $y \in R(x)$ we have $y \vDash \varphi$, and therefore $y \vDash \varphi \vee \psi$; since this holds for all $y \in R(x)$, it follows that $x \vDash \Box(\varphi \vee \psi)$.

(b) Observe that $(M,x) \vDash \Box(\varphi \wedge \psi)$ iff for all $y \in R(x)$ we have $(M,y) \vDash \varphi \wedge \psi$ iff for all $y \in R(x)$ we have $(M,y) \vDash \varphi$ and $(M,y) \vDash \psi$ iff $(M,x) \vDash \Box\varphi$ and $(M,x) \vDash \Box\psi$ iff $(M,x) \vDash \Box\varphi \wedge \Box\psi$.

(c) This is a bit of a trick question, since $\Diamond\varphi$ is just an abbreviation for $\neg\Box\neg\varphi$, so the task is to prove that $\vDash \neg\Box\neg\varphi \leftrightarrow \neg\Box\neg\varphi$, which is immediate.

(d) As in part (c), we use the fact that $\Diamond\varphi$ is an abbreviation for $\neg\Box\neg\varphi$, which means our task is to show that $\vDash \Box\varphi \leftrightarrow \neg\neg\Box\neg\neg\varphi$. So given any model M we must show that $[\![\Box\varphi]\!] = [\![\neg\neg\Box\neg\neg\varphi]\!]$. It is easy to see, first, that

$$[\![\neg\neg\Box\neg\neg\varphi]\!] = X \setminus (X \setminus [\![\Box\neg\neg\varphi]\!]) = [\![\Box\neg\neg\varphi]\!].$$

Now observe that $x \vDash \Box\neg\neg\varphi$ iff $R(x) \subseteq [\![\neg\neg\varphi]\!]$ iff $R(x) \subseteq [\![\varphi]\!]$, since $[\![\neg\neg\varphi]\!] = [\![\varphi]\!]$. Thus $[\![\Box\neg\neg\varphi]\!] = [\![\Box\varphi]\!]$, and we are done.   □

[10] Pictured here.

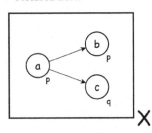

Exercise 73

Consider the model $M = (X, R, v)$ where $X = \{a,b,c\}$, $R = \{(a,b),(a,c)\}$, $v(p) = \{a,b\}$, and $v(q) = \{c\}$.[10] Observe that $(M,a) \vDash \Box(p \vee q)$, but $(M,a) \nvDash \Box p$ and $(M,a) \nvDash \Box q$, which refutes (a). Similarly, we can see that $(M,a) \vDash \Diamond p \wedge \Diamond q$ but $(M,a) \nvDash \Diamond(p \wedge q)$, so $(M,a) \nvDash (\Diamond p \wedge \Diamond q) \rightarrow \Diamond(p \wedge q)$. Finally, it is not hard to check that

$(M, a) \nvDash p \rightarrow \Box\Diamond p$, and $(M, a) \nvDash \Diamond\Box p \rightarrow \Box\Diamond p$ (remember that $\Box\varphi$ is vacuously true at any world x with $R(x) = \varnothing$, and $\Diamond\varphi$ is false at any such world).   □

Exercise 74

It is easy to see that in a trivial model, every formula of the form $\Box\varphi$ is valid, while every formula of the form $\Diamond\varphi$ is refuted at every state. It follows immediately that (b) and (c) are valid on the class of all trivial frames, since each is a material conditional with a valid consequent. Moreover, (a) is valid on the class of all trivial frames since each side of the biconditional is false at all states.

One can check that (a) is valid on the class of all frames where each world has exactly one accessible world; (b) is valid on the class of all frames where the relation is symmetric; and (c) is valid on the class of all frames where, for all w, whenever x, y are accessible from w, there exists a world z accessible from both x and y. (This is sometimes called the class of *convergent* frames.)

Let's prove this last claim in more detail. Let M be a model with the stated property, and w a point therein. Suppose $w \vDash \Diamond\Box\varphi$; then there must exist an x such wRx and $x \vDash \Box\varphi$. Now take any y with wRy; we want to show that $y \vDash \Diamond\varphi$ (since this will establish that $w \vDash \Box\Diamond\varphi$). By the given property, there is a z such that xRz and also yRz. Since xRz and $x \vDash \Box\varphi$, we must have $z \vDash \varphi$, which shows that $y \vDash \Diamond\varphi$, as desired.   □

Exercise 80

(a) The formula $\Box p \rightarrow \Diamond p$ defines the class of *serial* frames (i.e., frames such that $\forall x \exists y(xRy)$). It is easy to see that this formula is valid in any such frame. Conversely, given a frame $F = (X, R)$ and a point $x \in X$ such that $R(x) = \varnothing$, we have (for any model based on F) $x \vDash \Box p$ but $x \nvDash \Diamond p$.

(b) The formula $\Box(\Box p \rightarrow p)$ defines the class of *shift-reflexive* frames (i.e., frames such that $\forall x \forall y(xRy$ implies $yRy))$. To see this, first let M be shift-reflexive and suppose $x \in X$. Then for any $y \in R(x)$ we know yRy, from which it follows that $y \vDash$

$\Box p \to p$, hence $x \vDash \Box(\Box p \to p)$, as desired. Conversely, suppose that (X, R) is a frame that is not shift-reflexive, so there exist x, y such that xRy but not yRy. Define $v(p) = R(y)$. Then clearly $y \vDash \Box p$ but $y \nvDash p$, so $y \nvDash \Box p \to p$, so $x \nvDash \Box(\Box p \to p)$, as desired.

(c) The formula $\Diamond \Box \varphi \to \Box \Diamond \varphi$ defines the class of frames (X, R) such that, for all $w, x, y \in X$, if wRx and wRy then there exists a $z \in X$ such that xRz and yRz (i.e., the class of *convergent* frames). One direction of this was already established in the solution to Exercise 74.

For the other direction, suppose that (X, R) is a frame in which wRx and wRy but there is no z such that both xRz and yRz. Define $v(p) = R(x)$. Then certainly $x \vDash \Box p$, so $w \vDash \Diamond \Box p$. On the other hand, for all $t \in X$, if yRt then $t \notin [\![p]\!]$ by assumption, which means that $y \nvDash \Diamond p$. Thus, $w \nvDash \Box \Diamond p$, which completes the proof. $\qquad\square$

Exercise 81

(a) The formula $\Box p \leftrightarrow p$ defines the class of frames where R is the identity. If $F \in \mathcal{F}_{id}$ then one easily checks that $F \vDash \Box p \leftrightarrow p$. Conversely, if R is not the identity, then either there is an x with no self-loop, or an x and y such that xRy and $x \ne y$. In the first case, set $v(p) = R(x)$; then $x \vDash \Box p$ but $x \nvDash p$. In the second case, set $v(p) = \{x\}$; then $x \vDash p$ but $x \nvDash \Box p$.

(b) The formula $p \to \Box \Diamond p$ defines the class of symmetric frames. Let M be symmetric and suppose $x \vDash p$; then for all $y \in X$, if xRy then yRx, in which case $y \vDash \Diamond p$; hence, $x \vDash \Box \Diamond p$. Conversely, suppose that in $F = (W, R)$ we have xRy but not yRx. Define $v(p) = \{x\}$. Then $x \vDash p$ but clearly $y \nvDash \Diamond p$, hence $x \nvDash \Box \Diamond p$.

[11] Negative introspection!

(c) The formula $\neg \Box p \to \Box \neg \Box p$[11] defines the class of Euclidean frames. For validity, observe that if M is Euclidean and $x \vDash \neg \Box p$ then there is some $y \in R(x)$ such that $y \nvDash p$. Now consider an arbitrary $z \in R(x)$; by the Euclidean property we have zRy, so $z \vDash \neg \Box p$; since this holds for all $z \in R(x)$, it follows that $x \vDash \Box \neg \Box p$, as desired.

For the converse, suppose that F is not Euclidean, so we have points x, y, z with xRy and xRz but not yRz. Set $v(p) = R(y)$.

Then clearly $y \models \Box p$ so $x \models \Diamond \Box p$ and thus $x \not\models \Box \neg \Box p$. On the other hand, since $z \notin R(y)$, we know that $z \not\models p$, so since xRz, $x \not\models \Box p$, that is, $x \models \neg \Box p$, as desired. $\square$

Exercise 82

(a) The formula $Kp \to Bp$ defines the class of frames where $R_B \subseteq R_K$. Let F be such a frame; then if $w \models Kp$, we know that for all x such that wR_Kx, $x \models p$. Now consider y such that wR_By; by assumption, wR_Ky and therefore $y \models p$; hence, $w \models Bp$. Conversely, let $F = (X, R)$ be a frame such that for some $w, x \in X$ we have wR_Bx but not wR_Kx. Define $v(p) = X \setminus \{x\}$. Then clearly $w \models Kp$ but $w \not\models Bp$.

(c) The formula $Bp \to BKp$ defines the class of frames where, whenever wR_Bx and xR_Ky, we have wR_By. Suppose that F is such a frame and that $w \models Bp$. Consider a point x such that wR_Bx, and observe that we must have $x \models Kp$, since if not then there is some y with xR_Ky and $y \not\models p$, which contradicts the fact that $w \models Bp$ since, by assumption, we must have wR_By. It follows that $w \models BKp$.

Conversely, suppose that $F = (X, R)$ is a frame such that for some $w, x, y \in X$ we have wR_Bx and xR_Ky but not wR_By. Define $v(p) = X \setminus \{y\}$. Then clearly $w \models Bp$, but $x \not\models Kp$, and so $w \not\models BKp$. $\square$

Exercise 83

The validities in part (a) follow immediately from the fact that $\Box^*$ is a standard box-type modality that is being interpreted with respect to a reflexive and transitive relation.

For (b), let M be an arbitrary model and x an arbitrary point therein. Suppose that $x \models \Box^*(\varphi \to \Box\varphi)$ and $x \models \varphi$. We wish to show that $x \models \Box^*\varphi$, or in other words, that $R^*(x) \subseteq [\![\varphi]\!]$. We will establish this by proving the following: for all k, if $x_1, \ldots, x_k$ is an R-path and $x_1 = x$, then $x_k \models \varphi$. The proof proceeds by induction.

For the base case, $k = 1$, we have $x_k = x_1 = x$, and so since $x \models \varphi$ we are done. Now suppose inductively the result holds for k, and let $x_1, \ldots, x_{k+1}$ be an R-path. Then also $x_1, \ldots, x_k$ is an R-path, so $x_k \in R^*(x)$. Since $x \vDash \Box^*(\varphi \to \Box\varphi)$, it follows that $x_k \vDash \varphi \to \Box\varphi$. Moreover, by the induction hypothesis, $x_k \vDash \varphi$. Thus $x_k \vDash \Box\varphi$, so since $x_k R x_{k+1}$, we must have $x_{k+1} \vDash \varphi$, as desired. □

Exercise 86

[12] Notice that there is no requirement for a bisimulation to "connect" to every point of each model: here, the points in $X \setminus R^*(x)$ are not connected to anything! A similar trick is useful in reproving Proposition 42.

For Proposition 44, we can define a bisimulation $\sim$ between M and M_x by setting[12]

$$\sim \; = \{(y,y) \; : \; y \in M_x\}.$$

The BASE clause of being a bisimulation is satisfied because for all $y \in R^*(x)$ and any $p \in \text{PROP}$ we have $y \in v(p)$ iff $y \in v_{R^*(x)}(p)$; the FORTH and BACK conditions follow immediately from the fact that for all $y \in R^*(x)$, $R(y) = R_{R^*(x)}(y)$. □

Exercise 87

For (a), observe that $-10 \sim -1$, and $-10 < -5$. If $\sim$ were a bisimulation, we should be able to find an n such that $-1 < n$ and $-5 \sim n$. But this is impossible since $n > -1$ implies that $n \geq 0$, and the definition of $\sim$ guarantees that a negative integer cannot be related to a non-negative integer.

For (b), observe that $-1 \vDash \Box\neg\Diamond p$, while $-2 \nvDash \Box\neg\Diamond p$. On the other hand, we know that $-1 \sim -2$; if $\sim$ were a bisimulation, by Theorem 45 this would imply that -1 and -2 have the same theory, a contradiction. □

Exercise 88

(a) Suppose this class of frames were definable, say by the formula φ, and consider the frames $F_1 = (\{x\}, \varnothing)$ and $F_2 = (\{y\}, \{(y,y)\})$. Then since F_1 does not contain a self-loop, some model M_1 based on F_1 must refute φ, so $(M_1, x) \nvDash \varphi$. But in that case, if we take M_2 to be any model based on F_2, by invariance under disjoint unions (Proposition 42) we must have $(M_1 \sqcup M_2, x) \nvDash \varphi$. On the other hand, since $M_1 \sqcup M_2$ clearly contains a self-loop (at y), we must also have $M_1 \sqcup M_2 \vDash \varphi$, a contradiction.

(b) Suppose this class of frames were definable, say by the formula φ. Then we must have $(\mathbb{Z}, <) \vDash \varphi$, since every point in this frame has a predecessor. Consider now the frame $(\mathbb{N}, <)$; and note that this is the generated subframe of $(\mathbb{Z}, <)$ at the point 0. However, it is not the case that every point $(\mathbb{N}, <)$ has a predecessor. It follows that some model $M = (\mathbb{N}, <, v)$ refutes φ, and therefore (by invariance under generated submodels, Proposition 44) there is a model $M' = (\mathbb{Z}, <, v')$ that refutes φ (simply define $v'(p) = v(p)$ for all p), contradicting our original assumption.

(c) Suppose this class of frames were definable, say by the formula φ, and consider the frames $F = (\{w\}, \varnothing)$ and

$$F' = (\{x, y, z\}, \{(x, y), (y, z), (z, x)\}).$$

Clearly F does not contain a triangle, so we must have $F \nVdash \varphi$. In particular, there is some model M based on F and some world w therein such that $(M, w) \nVdash \varphi$. On the other hand, let M' be a model based on F' and consider the disjoint union $M \sqcup M'$. Clearly $M \sqcup M'$ *does* contain a triangle, so we must have $M \sqcup M' \vDash \varphi$; however, by Proposition 42, we must also have $(M \sqcup M', w) \nVdash \varphi$, a contradiction.

(d) Suppose this class of frames were definable, say by the formula φ. Consider the frame

$$F = (\{x, y, z\}, \{(x, z), (y, z), (x, x), (y, y)\}).$$

Clearly F is not reverse Euclidean, so some model M based on F must refute φ. Consider the generated submodel M_x; this model is based on the frame $F_x = (\{x, z\}, \{(x, z), (x, x)\})$, which *is* reverse Euclidean, so $M_x \vDash \varphi$. This implies (by Proposition 44) that $(M, x) \vDash \varphi$ and $(M, z) \vDash \varphi$. Similarly, M_y is based on a frame that is reverse Euclidean, so $M_y \vDash \varphi$, thus $(M, y) \vDash \varphi$. We have therefore shown that $M \vDash \varphi$, a contradiction. $\square$

Exercise 90

It is easy to see that f is surjective. To check that it's a bounded morphism, the FORTH condition is trivial (since R is the complete relation), while the BACK condition follows easily from the fact that, for any integer n, there exist integers m_0, m_1, and m_2 greater than n such that $m_i/3$ has remainder i.

Theorem 47 tells us that a surjective, bounded morphisms allows us to "pull back" refutations: in this case, given any model $M = (F, v)$ that refutes φ, we pull back the valuation v by defining $v' = f^{-1} \circ v$, and appeal to bisimulation invariance to conclude that $(\mathbb{Z}, <, v') \nvDash \varphi$.

Now since F fails to have each of the four properties listed in part (b), F would refute any purported defining formula. But then by the above argument, $(\mathbb{Z}, <)$ must also refute any such formula, despite the fact that this frame has each of these four properties. Hence, none of the properties can be definable in the basic modal language. □

Exercise 91

(a) Consider the model $M = (\{s, t\}, \{(s, t)\}, v)$ where $v(p) = \varnothing$. Observe that $(M, t) \nvDash \blacktriangleleft p$ since $(M, s) \nvDash p$. On the other hand, in the generated submodel M_t, we can see that $(M_t, t) \vDash \blacktriangleleft p$ vacuously (since s does not exist in M_t). Thus, by Proposition 44, we conclude that the modality $\blacktriangleleft$ is not expressible in the basic modal language.

(c) Consider the models $M = (\{a\}, \{(a, a)\}, v)$ and

$$M' = (\{b, c\}, \{(b, c), (c, b)\}, v'),$$

where $v(p) = \{a\}$ and $v'(p) = \{b, c\}$. It is easy to check that the relation $\sim$ such that $a \sim b$ and $a \sim c$ is a bisimulation. But $(M, a) \vDash \circlearrowright$ while $(M, b) \nvDash \circlearrowright$, which (by Theorem 45) shows that $\circlearrowright$ is not expressible in the basic modal language. □

Exercise 92

If γ is a formula of propositional logic, $p_1, \ldots, p_n \in \text{PROP}$, and $\varphi_1, \ldots, \varphi_n \in \mathcal{L}_\square$, then we can recursively define the *substitution* of $\varphi_1, \ldots, \varphi_n$ for $p_1, \ldots, p_n$ in γ, denoted

$$\gamma[\varphi_1/p_1] \cdots [\varphi_n/p_n],$$

in the obvious way:

$$p[\varphi_1/p_1] \cdots [\varphi_n/p_n] = \begin{cases} \varphi_i & \text{if } p = p_i \\ p & \text{otherwise} \end{cases}$$

$$(\neg\varphi)[\varphi_1/p_1] \cdots [\varphi_n/p_n] = \neg(\varphi[\varphi_1/p_1] \cdots [\varphi_n/p_n])$$

$$(\varphi \wedge \psi)[\varphi_1/p_1] \cdots [\varphi_n/p_n] = \varphi[\varphi_1/p_1]$$
$$\cdots [\varphi_n/p_n] \wedge \psi[\varphi_1/p_1] \cdots [\varphi_n/p_n].$$

In this case we call $\gamma[\varphi_1/p_1] \cdots [\varphi_n/p_n]$ a substitution instance of γ. If γ is deducible in classical propositional logic, then it is easy to see that any substitution instance $\gamma[\varphi_1/p_1] \cdots [\varphi_n/p_n]$ of γ is deducible in K, simply by applying the same substitutions to each line of a propositional deduction of γ.[13]   □

[13] If you are uncertain of this, try proving it by induction on the length of the deduction.

Exercise 96

To show that a formula φ is *not* deducible from a given axiom system, it suffices to find a model that refutes φ but with respect to which the given axiom system is sound. Since K is sound with respect to the class of all models, for (a) we need merely exhibit any model that refutes $\Box\varphi \to \Diamond\varphi$. For instance, it is easily seen that any model with a world w that has no accessible worlds refutes this formula.

For (b), we first show that KD is sound with respect to the class of serial frames. Proving soundness in this case simply requires showing that $\Box\varphi \to \Diamond\varphi$ is valid on any such frame (which is easy) and then "rerunning" the soundness proof for K (Theorem 52). Having established soundness, we must exhibit a model in this class (i.e., based on a serial frame) that refutes $\Box\varphi \to \varphi$. And indeed, it is easy to see that the two-point model $(\{a, b\}, \{(a, b), (b, a)\}, v)$ where $v(p) = \{b\}$ refutes $\Box p \to p$ at a.

For (c), we exhibit a formal T-deduction.

φ_1	$=$	$\Box\varphi \to \varphi$	(T)
φ_2	$=$	$\Box\neg\varphi \to \neg\varphi$	(T)
φ_3	$=$	$(\Box\neg\varphi \to \neg\varphi) \to (\varphi \to \Diamond\varphi)$	(taut)
φ_4	$=$	$\varphi \to \Diamond\varphi$	(2,3 MP)

$$\varphi_5 = (\Box\varphi \to \varphi) \to ((\varphi \to \Diamond\varphi) \to (\Box\varphi \to \Diamond\varphi)) \quad \text{(taut)}$$
$$\varphi_6 = (\varphi \to \Diamond\varphi) \to (\Box\varphi \to \Diamond\varphi) \qquad\qquad\qquad (1,5 \text{ MP})$$
$$\varphi_7 = \Box\varphi \to \Diamond\varphi \qquad\qquad\qquad\qquad\qquad (4,6 \text{ MP})$$

For (d), it is not hard to show that S4 is sound with respect to the class of reflexive, transitive frames (as we have already seen, it suffices to show that both $\Box\varphi \to \varphi$ and $\Box\varphi \to \Box\Box\varphi$ are valid on all such frames). We can then establish the result by exhibiting a reflexive, transitive model that refutes $\Diamond\Box\varphi \to \Box\varphi$; for instance, the model $(\{a,b\}, \{(a,b),(a,a),(b,b)\}, v)$ where $v(p) = \{b\}$ is reflexive and transitive but $a \nVdash \Diamond\Box p \to \Box p$.

Finally, for (e), we exhibit an S5-deduction (condensing some steps for brevity).

$$\varphi_1 = \Diamond\Box\neg\varphi \to \Box\neg\varphi \qquad\qquad\qquad\qquad (5)$$
$$\varphi_2 = \Box\neg\varphi \to \neg\varphi \qquad\qquad\qquad\qquad\qquad (T)$$
$$\varphi_3 = (\Diamond\Box\neg\varphi \to \Box\neg\varphi) \to ((\Box\neg\varphi \to \neg\varphi)$$
$$\qquad \to (\Diamond\Box\neg\varphi \to \neg\varphi)) \qquad\qquad\qquad \text{(taut)}$$
$$\varphi_4 = (\Box\neg\varphi \to \neg\varphi) \to (\Diamond\Box\neg\varphi \to \neg\varphi) \quad (1,3 \text{ MP})$$
$$\varphi_5 = \Diamond\Box\neg\varphi \to \neg\varphi \qquad\qquad\qquad\qquad (2,4 \text{ MP})$$
$$\varphi_6 = (\Diamond\Box\neg\varphi \to \neg\varphi) \to (\varphi \to \Box\Diamond\varphi) \quad \text{(taut)}$$
$$\varphi_7 = \varphi \to \Box\Diamond\varphi \qquad\qquad\qquad\qquad\qquad (5,6 \text{ MP})$$

$\square$

Exercise 102

(a) We must show that the canonical model built from S4 is both reflexive and transitive. Reflexivity follows just as it did in Proposition 60, since S4 includes the (T) axiom. So we just need to show that if $x_\Sigma \tilde{R} x_\Gamma$ and $x_\Gamma \tilde{R} x_\Delta$, then $x_\Sigma \tilde{R} x_\Delta$. Let $\Box\varphi \in \Sigma$; it suffices to show that $\varphi \in \Delta$. By (4), we know that $\Box\Box\varphi \in \Sigma$ as well, so since $x_\Sigma \tilde{R} x_\Gamma$ we must have $\Box\varphi \in \Gamma$, and then from $x_\Gamma \tilde{R} x_\Delta$ it follows that $\varphi \in \Delta$, as desired.

(b) It suffices to show that the canonical model based on K5 is Euclidean. So suppose that $x_\Sigma \tilde{R} x_\Gamma$ and $x_\Sigma \tilde{R} x_\Delta$; we want to show that $x_\Gamma \tilde{R} x_\Delta$. Suppose $\Box\varphi \in \Gamma$; then $\neg\Box\varphi \notin \Gamma$, so $\Box\neg\Box\varphi \notin \Sigma$ (since $x_\Sigma \tilde{R} x_\Gamma$). This implies that $\neg\Box\varphi \notin \Sigma$, since otherwise the (5) axiom would force $\Box\neg\Box\varphi \in \Sigma$, a contradiction. But then $\Box\varphi \in \Sigma$, so $\varphi \in \Delta$ (since $x_\Sigma \tilde{R} x_\Delta$) and we are done.

(c) Here the main trick is to show that a relation is an equivalence relation (reflexive, symmetric, and transitive) if and only if it is reflexive, transitive, and Euclidean.

(d) We must show that the canonical model built from KD is serial. Let $x_\Sigma \in \tilde{W}$, and observe that $\vdash_{KD}\Box(p \vee \neg p)$. Using (D), it follows that $\vdash_{KD}\Diamond(p \vee \neg p)$, and so by the Existence Lemma (Lemma 59), there is some $x_\Gamma \in \tilde{W}$ with $x_\Sigma \tilde{R} x_\Gamma$, which establishes seriality.

(e) This follows easily from previous proofs.

(f) Consider the canonical model based on K1.1. Suppose that $x_\Sigma \tilde{R} x_\Gamma$ and $x_\Sigma \tilde{R} x_\Delta$; we want to show that $x_\Gamma = x_\Delta$, or equivalently, that $\Gamma = \Delta$. So let $\varphi \in \Gamma$. Then $\neg\varphi \notin \Gamma$, from which it follows that $\Box\neg\varphi \notin \Sigma$, hence $\Diamond\varphi \in \Sigma$. But now by the additional axiom scheme, we must have $\Box\varphi \in \Sigma$, and so by definition of $\tilde{R}$, we can conclude that $\varphi \in \Delta$. This shows that $\Gamma \subseteq \Delta$; the reverse containment is completely analogous. $\square$

Exercise 105

For (a), it suffices to observe that whenever $(M, x) \nvDash \varphi$, by invariance under generated submodels we also have $(M_x, x) \nvDash \varphi$, and M_x is rooted (with root x).

For (b), take any φ such that $\nvdash_{S5} \varphi$. We must show that φ is refuted on some frame $F = (X, R)$ in which $R = X \times X$. We produce such a frame now. Let $M = (X, R, v)$ be a model in which the relation is an equivalence relation and such that $(M, w) \nvDash \varphi$. This must exist by Exercise 102(a). Now let $M_w = (X_w, R_w, v_w)$ be the submodel of M generated by w. It is easy to check that $X_w = R(w)$ and that $R_w = X_w \times X_w$; this is because R is an equivalence relation so the submodel generated by a point is just the equivalence class of that point. Now by invariance under generated submodels, we can conclude that $(M_w, w) \nvDash \varphi$, as desired.

For (c), as above we consider a submodel M_w generated by w, where w refutes a given non-theorem of KD45 and M is a serial, transitive, and Euclidean model (cf. Exercise 102(c)). Then it is straightforward to check that M_w is a brush—it has final cluster equal to X_w if wRw, and otherwise it has final cluster equal to $R(w)$. $\square$

Exercise 109

(a) R^+ contains R because if $(x, y) \in R$ then y is R-reachable from x via the sequence (x, y). R^+ is transitive because if y is R-reachable from x via the sequence $z_1, \ldots, z_k$ and z is R-reachable from y via the sequence $z'_1, \ldots, z'_\ell$, then z is R-reachable from x via the sequence $z_1, \ldots, z_k, z'_2, \ldots, z'_\ell$.

To see that R^+ is the smallest such relation, let S be an arbitrary transitive relation containing R. We'll show that $R^+ \subseteq S$. Indeed, if $(x, y) \in R^+$, then y is R-reachable from x, and since S contains R, we know that y is also S-reachable from x. But because S is transitive, it is easy to see that this implies that $(x, y) \in S$, as desired.

(b) Consider the frame $(\{x, y\}, \{(x, y), (y, x)\})$, and let $(\vec{X}, \vec{R})$ denote its unravelling about x. Observe that $(x)\vec{R}^+(x, y, x)$ (since $(x)\vec{R}(x, y)$ and $(x, y)\vec{R}(x, y, x)$) and $f(x) = f(x, y, x) = x$, but it is not the case that xRx. This shows that f is not a bounded morphism from $(\vec{X}, \vec{R}^+)$ to (X, R).

Now suppose that (X, R) is a transitive frame. To show that f is a bounded morphism from $(\vec{X}, \vec{R}^+)$ to (X, R), we must prove that the BACK and FORTH conditions are satisfied.

First suppose that $f(\vec{w})Rx$; we need to find a $\vec{y} \in \vec{X}$ such that $\vec{w}\vec{R}^+\vec{y}$ and $f(\vec{y}) = x$ (this is the BACK condition). But this is easy (and identical to the original proof): let $\vec{y} = \vec{w}^\frown x$ (i.e., the sequence obtained by concatenating x to the end of $\vec{w}$). Then, by definition, $\vec{w}\vec{R}\vec{y}$ (so of course $\vec{w}\vec{R}^+\vec{y}$ as well), and likewise $f(\vec{y}) = x$.

For the FORTH condition, suppose that $\vec{w}\vec{R}^+\vec{y}$; we must show that $f(\vec{w})Rf(\vec{y})$. Since $\vec{w}\vec{R}^+\vec{y}$, it follows from part (a) that there exist $x_1, \ldots, x_k$ such that $\vec{y} = \vec{w}^\frown(x_1, \ldots, x_k)$. In particular, we have $f(\vec{y}) = x_k$, $f(\vec{w})Rx_1$, and $x_i Rx_{i+1}$ for each $1 \le i < k$. But since R is transitive, it follows that $f(\vec{w})Rx_k$, which yields $f(\vec{w})Rf(\vec{y})$, as desired.

(c) Suppose that $\nvdash_{K4} \varphi$. Since K4 is complete with respect to the class of transitive frames, we can find a transitive model M that refutes φ at some point w. Let M_w be the generated submodel of M at w; let $\vec{M}_w$ denote its unravelling about w; let $\vec{M}_w^+$ denote the transitive closure of this model. Clearly, this model is based on a t-tree; moreover, since each of the operations invoked in its construction has been shown to preserve the truth of modal formulas, we have $(\vec{M}_w^+, w) \nvDash \varphi$.

(d) Let $\bot$ denote a propositional contradiction. The "leaves" of t-trees have no outgoing edges; any such node satisfies $\Box\bot$. Since every point in a finite t-tree that is not a leaf can "see" a leaf (i.e., there is an edge from every non-leaf to some leaf), it follows that every point in a finite t-tree satisfies $\Box\bot \lor \Diamond\Box\bot$. However, it is easy to check that this formula is not a theorem of K4. $\qquad\qquad\qquad\qquad\qquad\Box$

Exercise 110

As in the unravelling construction (Proposition 62), it suffices to build, for each frame F, an *irreflexive* frame F' with a surjective bounded morphism $F : F' \to F$. Using such a construction, completeness follows easily: let φ be a non-theorem; find an arbitrary frame F that refutes it; observe that the bounded morphism $f : F' \to F$ allows us to "pull back" this refutation to F'; conclude that every non-theorem is refuted on an irreflexive frame.

We now provide an instance of the desired construction. Given $F = (X, R)$, set $X' = X \times \{1, 2\}$ and define $f : X' \to X$ by

$$f(x, 1) = f(x, 2) = x.$$

In effect, we have "doubled" the original state space X, and f is just the function that collapses each pair of "twins" back into one. Next, define R' by

$$(x, i)R'(y, j) \text{ iff } xRy \text{ and } i \neq j.$$

Intuitively, R' behaves just like R except it "jumps back and forth" between the two halves of the "doubled" state space X'— in other words, it only puts edges between points that lie in different halves of X'. As such, it is easy to see that R' is irreflexive. It is also easy to check that f is a bounded morphism from F' to F. Since f is clearly surjective, we are done. $\qquad\square$

Exercise 112

For (a), suppose that M is reflexive. Let $|w|_\Sigma \in X_\Sigma$. Recall that $|w|_\Sigma R_\Sigma |w|_\Sigma$ just in case there exist $x, y \in |w|_\Sigma$ such that xRy. But this is trivially true, since $w \in |w|_\Sigma$ and wRw by assumption.

Now suppose that $\nvdash_T \varphi$. Then φ is refuted on some reflexive model M (by Corollary 61). It follows that φ is refuted on M_{Σ_φ} (where Σ_φ is the set of subformulas of φ), which is both finite and reflexive. Hence T is complete with respect to the class of finite reflexive frames.

For (b), consider the model $M = (\{w, x, x', y\}, \{(w, x), (x', y)\}, v)$, where $v(p) = \{w\}$ and $v(q) = \{y\}$. Clearly M is transitive. Let $\Sigma = \{p, q\}$; then it is easy to see that $W_\Sigma = \{\{w\}, \{x, x'\}, \{y\}\}$ and $R_\Sigma = \{(\{w\}, \{x, x'\}), (\{x, x'\}, \{y\})\}$, so M_Σ is clearly *not* transitive. $\qquad\square$

Exercise 113

(a) Suppose that $\nvdash_K \varphi$. By Theorem 66, we can find a finite model $M = (X, R, v)$ that refutes φ at some point $x \in X$. Without loss of generality we may assume that M is rooted with root x (passing to a generated submodel if necessary). Note that because M is finite, it has bounded degree (with bound $|X|$). Let $\vec{M}$ be the unravelling of M about x. Then we know that $\vec{M}$ is based on a tree; moreover, the unravelling construction guarantees that for every point $(r, x_1, \ldots, x_k) \in \vec{X}$, we have $|\vec{R}(r, x_1, \ldots, x_k)| = |R(x_k)|$. It follows that $\vec{M}$ has bounded degree as well; since $(\vec{M}, \vec{x}) \nvDash \varphi$, this completes the proof.

(b) Suppose that $\nvdash_{KD} \varphi$; then by Exercise 102(b), there is a serial model $M = (X, R, v)$ that refutes φ at some point $x \in X$. Let $y \in R(x)$, and consider the model $M' = (X', R', v')$ where

- $X' = X \sqcup \{y_k : k \in \mathbb{N}\}$;
- $R' = R \cup \{(z, y_k) : zRy, k \in \mathbb{N}\} \cup \{(y_k, z) : yRz, k \in \mathbb{N}\};$

- $v'(p) = \begin{cases} v(p) & \text{if } y \notin v(p) \\ v(p) \cup \{y_k : k \in \mathbb{N}\} & \text{if } y \in v(p). \end{cases}$

Then it is easy to check that the function $g : X' \to X$ defined by

$$g(w) = \begin{cases} w & \text{if } w \in X \\ y & \text{if } w = y_k \end{cases}$$

is a bisimulation, so $(M', x) \nVDash \varphi$, and by construction we have $|R'(x)| = \infty$ (since for all $k \in \mathbb{N}$, $y_k \in R(x)$).

This construction relied on the fact that $R(x) \neq \varnothing$. Note that $\nvdash_K \neg\Box\bot$, yet any point of infinite degree (indeed, any point with *any* outgoing edges) satisfies $\neg\Box\bot$. $\qquad\square$

Exercise 118

For (a), first show that $\vdash_{T_nC} C\varphi \to E(\varphi \wedge C\varphi)$, and then apply the induction rule. For (b), we can proceed by induction on k. The base case is just (C1). Now if we assume inductively that $\vdash_{T_nC} C\varphi \to E^k\varphi$, then by K_i-necessitation and K_i-distribution (applied for each K_i), we obtain $\vdash_{S5_nC} EC\varphi \to EE^k\varphi$. Putting this together with (C2) yields the desired result. $\qquad\square$

Exercise 123

To show that $int(X) = X$, by Proposition 71, it's enough to show that $X \subseteq int(X)$. So let $x \in X$; then by (T1) there is some $U \in \mathcal{T}$ such that $x \in U$; clearly also $U \subseteq X$, which establishes that $x \in int(X)$, as desired.

To show that $int(A \cap B) = int(A) \cap int(B)$, first suppose that $x \in int(A \cap B)$. Then there is an open set U such that $x \in U \subseteq A \cap B$. Of course, U then also witnesses that $x \in int(A)$ and $x \in int(B)$, proving containment. For the reverse containment, suppose that $x \in int(A) \cap int(B)$. Then there are open sets U and V such that $x \in U \subseteq A$ and $x \in V \subseteq B$. It follows immediately that $x \in U \cap V \subseteq A \cap B$; by (T2), the intersection of open sets is open, so this shows that $x \in int(A \cap B)$, as desired. $\qquad\square$

Exercise 129

It suffices to show that the axiom $\neg\Box\varphi \to \Box\neg\Box\varphi$ is not valid on all topological models. Consider the real line $\mathbb{R}$ with its usual topology, and let $v(p) = (0,1)$. Then $0 \models \neg\Box p$, but $0 \not\models \Box\neg\Box p$. $\qquad\square$

Exercise 130

(a) $\varphi \to \Box\varphi$ defines the class of spaces where every set is open. To see this, first suppose that $(X, \mathcal{T})$ is such a space (i.e., $\mathcal{T} = 2^X$). It follows that for every A, $int(A) = A$, and so in particular $A \subseteq int(A)$. This in turn shows that $\varphi \to \Box\varphi$ is valid on $(X, \mathcal{T})$, as desired.

   Conversely, suppose that *not* every set is open in some topological space $(X, \mathcal{T})$. Then there is some set A such that $int(A) \subsetneq A$. Let $x \in A \setminus int(A)$, and define $[\![p]\!] = A$. Then clearly $x \models p$ but $x \not\models \Box p$, hence $x \not\models p \to \Box p$, so this formula is not valid on $(X, \mathcal{T})$.

(b) The formula $\Diamond\varphi \to \Box\Diamond\varphi$ defines the class of spaces where every closed set is open. To see this, first suppose that $(X, \mathcal{T})$ is such a space. Then since $[\![\Diamond\varphi]\!] = cl([\![\varphi]\!])$, we know that $[\![\Diamond\varphi]\!]$ is closed, and therefore open. Thus $int([\![\Diamond\varphi]\!]) = [\![\Diamond\varphi]\!]$, so in particular $[\![\Diamond\varphi]\!] \subseteq int([\![\Diamond\varphi]\!]) = [\![\Box\Diamond\varphi]\!]$. This shows that $\Diamond\varphi \to \Box\Diamond\varphi$ is valid on $(X, \mathcal{T})$, as desired.

   Conversely, suppose that *not* every closed set is open in some topological space $(X, \mathcal{T})$. Then there is some closed set A such that $int(A) \subsetneq A$. Let $x \in A \setminus int(A)$, and define $[\![p]\!] = A$. Since A is closed, we also have $A = [\![\Diamond p]\!]$. It follows that $x \models \Diamond p$; on the other hand, by assumption $x \notin int([\![\Diamond p]\!])$, so $x \not\models \Box\Diamond p$. Thus $x \not\models \Diamond p \to \Box\Diamond p$, so this formula is not valid on $(X, \mathcal{T})$. $\qquad\square$

Exercise 132

First let $(X, \mathcal{T})$ be an Alexandroff space and take $x \in X$. Let $\mathcal{N}_x = \{U \in \mathcal{T} : x \in U\}$ and define

$$U_x = \bigcap_{U \in \mathcal{N}_x} U.$$

We know that $U_x \in \mathcal{T}$ because it is an intersection of open sets. We also know that U_x is the minimal neighbourhood of x because if U is any other open set containing x, then it occurs in the intersection defining U_x.

Conversely, suppose that every $x \in X$ has a minimal neighbourhood U_x. Let $\{U_\alpha : \alpha \in A\}$ be an arbitrary collection of open sets, and consider

$$U = \bigcap_{\alpha \in A} U_\alpha.$$

We wish to show that U is open. So let $x \in U$. It follows that for every α, $x \in U_\alpha$. By minimality, then, we know that for every α, $U_x \subseteq U_\alpha$, and therefore $U_\alpha \subseteq U$. Therefore we have shown that $x \in U_x \subseteq U$, so $x \in int(U)$. Since x was chosen arbitrarily from U, it follows that $U \subseteq int(U)$, so U is open. $\qquad \square$

Exercise 133

(a) $R_\mathcal{T}$ is reflexive because (trivially) $x \in cl(\{x\})$. Now suppose that $x R_\mathcal{T} y$ and $y R_\mathcal{T} z$, which means that $x \in cl(\{y\})$ and $y \in cl(\{z\})$. Consider any open set U containing x. Since $x \in cl(\{y\})$, we must also have $y \in U$; thus, since $y \in cl(\{z\})$, we can deduce that $z \in U$. We have therefore shown that every open set containing x contains z, or in other words, $x \in cl(\{z\})$; this implies $x R_\mathcal{T} z$ and proves transitivity.

(b) $R_\mathcal{T}$ need not be symmetric. Consider the set $X = \{x, y\}$ equipped with the topology $\mathcal{T} = \{X, \varnothing, \{y\}\}$. It is easy to see that $x \in cl(\{y\})$ but $y \notin cl(\{x\})$.

(c) We prove this by induction on φ. Since the base cases and the inductive steps corresponding to the Boolean connectives are trivial, we skip straight to the inductive step corresponding to $\square$. So suppose inductively that the result holds for φ; we will prove it holds for $\square\varphi$ as well.

First suppose that $((X, \mathcal{T}, v), x) \vDash \square\varphi$. Then there is an open set $U \in \mathcal{T}$ such that $x \in U \subseteq \llbracket \varphi \rrbracket$. Observe that for any y such that $x \in cl(\{y\})$, by definition we must have $y \in U$. It follows that for any such y, $((X, \mathcal{T}, v), y) \vDash \varphi$. By the inductive hypothesis, we also have $((X, R_\mathcal{T}, v), y) \vDash \varphi$. But this implies that $((X, R_T, v), x) \models \Box\varphi$, since the set of y we considered above consists of precisely those points accessible from x via R_T.

Conversely, suppose that $((X, R_T, v), x) \models \Box\varphi$. Let U denote the set of all y such that $x \in cl(\{y\})$; we then have, for all $y \in U$, $((X, R_T, v), y) \models \varphi$, and therefore by the inductive hypothesis $((X, T, v), y) \models \varphi$ as well. To show that $((X, T, v), x) \models \Box\varphi$ it therefore suffices to show that $U \in T$. In fact, we will show that $U = U_x$, the minimal neighbourhood of x (which is guaranteed to exists because X is Alexandroff).

First observe that if $x \in cl(\{y\})$ then, by definition, $y \in U_x$. This shows that $U \subseteq U_x$. On the other hand, if $x \notin cl(\{y\})$, then there is some open set V such that $x \in V$ but $y \notin V$. Since U_x is minimal, we have $U_x \subseteq V$ and therefore $y \notin U_x$. This shows (in contrapositive) that $U_x \subseteq U$, completing the proof.

(d) An example here is helpful. Consider the real line $\mathbb{R}$ with its usual topology $T_\mathbb{R}$, and observe that the induced relation $R_{T_\mathbb{R}}$ is just the identity relation (since no point of $\mathbb{R}$ is in the closure of any other point besides itself). Now consider a valuation on $\mathbb{R}$ where $v(p) = \{x \in \mathbb{R} \,:\, x > 0\}$. Then it is easy to see that $((\mathbb{R}, T_\mathbb{R}, v), 0) \models \Diamond p$ (since 0 is in the closure of the positive numbers), while $((\mathbb{R}, R_{T_\mathbb{R}}, v), 0) \not\models \Diamond p$ (since 0 is only accessible to itself, and $0 \models \neg p$). □

Exercise 136

First observe that we do certainly have $A \in 2^A$, since A is always a subset of itself. However, for $A \subseteq 2^A$ to hold, it would have to be the case that for each $x \in A$, we also have $x \in 2^A$. But consider the example given in the text: although $2 \in \{2, 4, 8\}$, it is *not* true that $2 \in 2^{\{2,4,8\}}$. It *is* true that $\{2\} \in 2^{\{2,4,8\}}$, but the set containing 2 is not the same mathematical object as 2 itself! □

Exercise 139

To see that R_A is not transitive, consider the words "cat", "top", and "pen". To see that it is not reflexive consider the word "cat". To see that it is not acyclic, consider the word "pop".

R_B is not reflexive since, for example, $|2(2) - 2| = 2 \not< 1$, so it is not the case that $2R_B2$. But it is not irreflexive either, since, for example, $|2(0) - 0| = 0 < 1$, so $0R_B0$. And it is not symmetric since $1R_B2$ but not $2R_B1$, as is easily checked.

For R_C, first observe that $3x + y$ is even just in case $x + y$ is even.[14] This makes it clear that R_C is symmetric (since $x + y = y + x$). It also must be transitive: suppose that xR_Cy and yR_Cz; then $x + y$ and $y + z$ are even, so also $x + 2y + z$ is even, and thus $x + z$ is even. Finally, R_C is not negatively transitive, since for example $1 + 2$ is odd, and $2 + 3$ is odd, but $1 + 3$ is even.   □

[14] Because $3x + y = 2x + (x + y)$.

Exercise 142

If $X = \{x, y\}$ and $R = X \times X$, then R is (trivially) negatively transitive, but not asymmetric. If $X = \{x, y, z\}$ and $R = \{(x, y)\}$, then R is obviously asymmetric, but it is not negatively transitive since $(x, z) \notin R$ and $(z, y) \notin R$, but $(x, y) \in R$. Thus asymmetry and negative transitivity are independent.

Note that in the previous example R is also transitive, so this shows that transitivity does not imply negative transitivity. Moreover, if $X = \{x, y, z\}$ and $R = \{(x, y), (y, x), (y, z), (z, y)\}$, then it is easy to check that R is negatively transitive, but it clearly fails to be transitive. Thus transitivity and negative transitivity are independent.

For part (c), first let's observe that irreflexivity and negative transitivity are independent. If $X = \{x\}$ and $R = \{(x, x)\}$, then clearly R is not irreflexive, but it is trivially negatively transitive. On the other hand, in the earlier example where R is asymmetric but not negatively transitive, clearly also R is irreflexive. Thus we have shown that irreflexivity and negative transitivity are independent, and we've already seen that negative transitivity and asymmetry are independent; on the other hand, we know that asymmetry implies irreflexivity, so certainly irreflexivity and asymmetry are not independent. Thus independence is not transitive.   □

Exercise 146

Part (a) is an immediate consequence of the fact that E is reflexive. For part (b), assume first that xEy and let $z \in [x]_E$. Then zEx, so by transitivity zEy, which means that $z \in [y]_E$. On the other hand if $z \in [y]_E$, then zEy, so by symmetry yEz, so by transitivity xEz, so by symmetry zEx, and thus $z \in [x]_E$. This shows that $[x]_E = [y]_E$. Now assume that $[x]_E = [y]_E$. Then, since $x \in [x]_E$ (by part (a)), by assumption we must have $x \in [y]_E$, which means xEy.

For (c), we'll prove the contrapositive. Suppose that $z \in [x]_E \cap [y]_E$. Then zEx and zEy, so by symmetry xEz, and by transitivity xEy. Part (b) then allows us to conclude that $[x]_E = [y]_E$, as desired. $\square$

Exercise 150

First we show that for all $n \geq 5$, $n! > 2^n$. For the base case $n = 5$ we have: $5! = 120 > 32 = 2^5$. Now suppose inductively the result holds for n. We want to show it holds for $n + 1$. Observe that:

$$(n + 1)! = (n + 1) \cdot n! > (n + 1) \cdot 2^n > 2 \cdot 2^n = 2^{n+1},$$

so we are done.

Next we show that for all $n \geq 5$, $2^n > n^2$. For the base case $n = 5$ we have: $2^5 = 32 > 25 = 5^2$. Now suppose inductively the result holds for n. We want to show it holds for $n + 1$. We have $2^{n+1} = 2^n + 2^n$ and $(n + 1)^2 = n^2 + 2n + 1$. By the inductive hypothesis, we also have $2^n > n^2$, so we would be done if we also knew that $2^n \geq 2n + 1$. We can also prove this by induction!

Let's use a different variable to keep things organized: we'll show that for all $m \geq 5$, $2^m \geq 2m + 1$. For the base case $m = 5$ we have $2^5 = 32 \geq 11 = 2(5) + 1$. Now suppose inductively the result holds for m. We want to show it holds for $m + 1$. We have $2^{m+1} = 2^m + 2^m$ and $2(m + 1) + 1 = 2m + 1 + 2$; since $2^m \geq 2m + 1$ and $2^m > 2$, we are done. $\square$

Exercise 154

For part (c): consider cutting the $2^n \times 2^n$ board into four quadrants.

Exercise 155

This is one of those great questions where, once you've found the correct solution, you *know it*. So if you're here because you *think* you found the solution but you're not sure, then you didn't.[15]

[15] Hint: there is absolutely nothing wrong with the part of the argument where we assume inductively that for any herd of size n, all horses in that herd are the same colour. This is precisely how induction works! We try to show that *if* it's true for n, then it's also true for $n + 1$. Keep trying!

Index

Printed in the United States
by Baker & Taylor Publisher Services